Marwah Ezzulddin Merza Al-Abasy
Basil Shukr Mahmood

Modeling and FPGA Implementation of ANN Based Electronic Circuits

Marwah Ezzulddin Merza Al-Abasy
Basil Shukr Mahmood

Modeling and FPGA Implementation of ANN Based Electronic Circuits

Noor Publishing

Imprint

Any brand names and product names mentioned in this book are subject to trademark, brand or patent protection and are trademarks or registered trademarks of their respective holders. The use of brand names, product names, common names, trade names, product descriptions etc. even without a particular marking in this work is in no way to be construed to mean that such names may be regarded as unrestricted in respect of trademark and brand protection legislation and could thus be used by anyone.

Cover image: www.ingimage.com

Publisher:
Noor Publishing
is a trademark of
Dodo Books Indian Ocean Ltd., member of the OmniScriptum S.R.L Publishing group
str. A.Russo 15, of. 61, Chisinau-2068, Republic of Moldova Europe
Printed at: see last page
ISBN: 978-620-2-79109-0

Modeling and FPGA Implementation of ANN Based Electronic Circuits

By

Marwa Izz Al-Deen Merza

ACKNOWLEDGMENTS

First, my great thanks go to **ALLAH** who helped and gave me the ability to achieve this thesis.

I also wish to thank my advisor, Dr. Basel Shukr Mahmood, for supporting my research and providing me the opportunity to learn so much.

I like to thank Dr. Dia Mohmad Ali for his help to understand LabVIEW.

I am also thankful to the Mr.Hassan Rahhal, Engineer in National Instruments who helped me to understand Elvis.

Thanks are giving to Toka Abdul-Hameed for her help and guidance.

Also, I want to thank my family for always being so supportive specially my dear mother.

Last and not least I would like to extend my thanks to all my friends and every one helped and supported me during preparing this thesis.

Contents

CHAPTER FIVE
CONCLUSIONS AND FUTURE WORKS

References

List of Figures

Figure No.	Name of figure
Figure (2-1)	The Components of a Processing Unit
Figure (2-2)	Neural Network Classification

Figure (4-24b)	First Recurrent layer in LabVIEW FPGA VI looping the design FPGA VI with less resources
Figure (4-25)	VI for First recurrent layer
Figure (4-26)	VI for Second Recurrent Layer
Figure (4-27)	Output Layer Of Elman Neural Network
Figure (4-28)	The mathematics List in LabVIEW
Figure (4-29)	Back-propagation of error in LabVIEW
Figure (4-30a)	Many inputs waveforms in LabVIEW training signals (patterns)
Figure (4-30b)	Many inputs waveforms in LabVIEW desired output after training
Figure (4-30c)	Output signal of RC circuit
Figure (4-31)	Parallel processing of Array element in LabVIEW FPGA
Figure (4-32)	Sequential Processing of Array elements in LabVIEW FPGA with less resources
Figure (4-33)	First recurrent layer
Figure (4-34)	First layer with memory
Figure (4-35)	Input and first layer sub VI
Figure (4-36)	Second layer
Figure (4-37)	Second layer with memory

List of Abbreviations

ALS	Adaptive Local Search
ADE	Application-specific Development Environment
ANN	Artificial Neural Network
DSP	Digital Signal Processing
DSI	Dynamic Systems Identification
ELVIS	Educational Laboratory Virtual Instrumentation Suite
ENN	Elman Neural Network
FPGA	Field Programmable Gate Array
FRNN	Fully Recurrent Neural Network
G	graphical language
GUI	Graphical User Interface
I/O	Input /Output
ISE	Integrated Software Environment
LUT	Look-Up Tables
MSE	Mean Square Error
NI	National Instruments
PID	Positive Adjustable Regulator
BP	Back Propagation
SRN	Simple Recurrent Network
SIT	Simulation Interface Toolkit
VHDL	VHSIC (Very High Speed Integrated Circuit) Hardware Description Language
VI	Virtual Instruments

List of Symbols

B	the slop of the nonlinear function $f(n)$
E	the error term
$f(n)$	approximation of tan-sigmoid function
g	the gain of the nonlinear function $f(n)$
WL1, WL2, Wo	the matrixes that contains the weights of feed forward connection
VL1, VL2	the matrixes that contains the weights of feedback connection
$y_d(t)$	desired output
$y_a(t)$	actual output
y_{-ink}	The results of the output unit
y_k	the result of the output unit after activation
z_{-inj}	the result of the hidden unit
z_j	the result of the hidden unit after activation
δ_k	delta of the output unit
δ_j	delta of the hidden unit

CHAPTER ONE

INTRODUCTION

1.1 <u>Overview:-</u>

Neural networks have shown a great progress in identification of nonlinear systems. There are certain characteristics in ANN which assist them in identifying complex nonlinear systems. ANNs are made up of many nonlinear elements and this gives them an advantage over linear techniques in modeling nonlinear systems. ANN are trained by adaptive learning, the network 'learns' how to do tasks, perform functions based on the data given for training. The knowledge learned during training is stored in the synaptic weights [1].

There is a variety of neural net architectures that have been developed over the past few decades. Beginning from the feed-forward networks and the multi-layer perceptrons (MLP) to the highly dynamic recurrent ANN, there are many different types of neural nets. Also, various algorithms have been developed for training these networks to perform the necessary functions [2]. LabVIEW has been proved in [2] as a very powerful software tool for building NN

NI LabVIEW is a complete graphical development environment used by thousands of engineers and scientists to efficiently design, prototype, and deploy embedded applications. LabVIEW combines hundreds of prewritten libraries, tight integration with off-the-shelf hardware, and a variety of programming approaches including graphical development, .m file scripts, and connectivity to existing C and HDL code. Whether designing medical devices or complex robots, one can reduce time to market and the overall cost of embedded design with LabVIEW and NI embedded hardware.

The NI LabVIEW FPGA Module extends LabVIEW graphical development to field-programmable gate arrays (FPGAs) on NI Reconfigurable I/O hardware. LabVIEW is distinctly suited for FPGA programming because it clearly represents parallelism and data flow. With the LabVIEW FPGA Module, custom measurement and control hardware can be created without low-level hardware description languages or board-level design. This custom hardware is used for unique timing and triggering routines, ultrahigh-speed control, interfacing to digital protocols, digital signal processing (DSP), RF and communications, and many other applications requiring high-speed hardware reliability and tight determinism.

In this thesis, a new approach has been applied to build neural network using Elvis system consisting of LabVIEW FPGA board that has been used to design a supervised recurrent neural network .In Elvis, the designer can use either a VHDL code or LabVIEW in LabVIEW FPGA environment to design neural network that can be implemented on FPGA. Here, LabVIEW FPGA has been used to design and implement recurrent and feed-forward neural network for electronic system.

1.2 <u>Literature Review</u>

In 1993, G. L. Dempsey, et al. [2] presented a paper that proposes the implementation of a neural-based 4-bit analog-to-digital converter and discussed many problems they encountered such as missing output codes.

In 1997, S.WhanLee and H.HeonSong [3] presented a paper proposed a new type of recurrent neural-network architecture, in which each output unit is connected to itself and is also fully connected to other output units and all hidden units. The proposed recurrent neural network differs from Jordan's and Elman's recurrent neural networks with respect to function and architecture, because it has been originally extended from being a mere multilayer feed-forward.

In 1999, E.O.Dijk [4] whose thesis presented a recurrent neural networks can be applied to the task of speaker independent phoneme recognition. Several recurrent neural network architectures found in literature are listed and categorized. A general modular description method is used to describe all the architectures found. The state-space neural network architecture and the Fully Recurrent Neural Network (FRNN) are investigated in more detail.

P.Sridhar, Sriram [5] in 2005 a thesis developed both supervised and the unsupervised neural nets successfully using LabVIEW, and it has been proved that LabVIEW is a very powerful software tool for building neural nets.

In 2006, a paper presented by A.Kalinli and S.Sagiroglu [6] new recurrent neural network (RNN) structure called ENEM for dynamic system identification. Affected by ENEM structure is based on Elman network and NARX neural network.

In 2008, Yuan-Chu Cheng, Wei-Min Qi, Jie Zhao[7] proposed a new modified Elman network. Then the dynamic characteristics of Elman series neural network are fully discussed. The theory study proved that the new Elman network inherently has proportional (P), integral (I), derivative (D) properties and traditional Elman networks only have PI or I characteristics.

In 2009, Z. Aimin and et al [8] presented a paper, two types of neural-network model have been developed for interferences in both frequency and time domain. For frequency domain interferences, the RNN with no time sequence input has been set up for internal circuits of electronic devices; and for interferences in time domain, the time-sequence-input-containing RNN model groups for internal circuits have been set up according to the characteristics of interference waveforms. This kind of model is applied when unsatisfying test results of the RNN model appear, which are caused by huge changes of the interference waveforms of internal circuits with no significant regularity.

In 2010, J.A.Samath, P.S.Kumar and A.Begum [9], introduced neural network algorithm to study the singular system of a linear electrical circuit for time invariant and time varying cases. Neural network or simply neural nets are computing systems, which can be trained to learn a complex relationship between two or many variables or data sets. Having the structures similar to their biological counterparts, neural networks are representational and computational models processing information in a parallel distributed fashion composed of interconnecting simple processing nodes.

In 2010, an Adaptive Local Search (ALS) algorithm was proposed by Z.Zhang [10] to train Elman Neural Network (ENN) for Dynamic Systems Identification (DSI) from a new angle instead of traditional Back-Propagation (BP) based gradient descent technique by Zhang, et al .

In the same year, a paper presented by T. Rashid [11] that focuses on a study of different Recurrent Neural Network architectures. A new alternative type of recurrent neural network is introduced. This alternative recurrent neural network is based on the simple recurrent network (SRN) but has an architecture that consists of fully interconnected layers. The network has feedback connections from the input, hidden, and output layers, with each connecting to its own designated context layer.

1.3   Aims of the Thesis

1- This thesis aims to use intelligent approach (artificial neural network) for modeling electronic circuits to make adjusting to the properties of electronic circuits.

2- Neural network is used to compute the relationship between the input and output of any electronic circuit and accordingly equivalent circuit of any electronic circuit (hardware component in any electronic circuit) can be realized by neural network.

3- The electronic circuits that designed with neural network are not affected by changes of temperature and external noise.

4- Build a FPGA hardware for the feature extraction system of that electronic circuit.

1.4   Thesis Organization

This thesis is consists of Five chapters. *Chapter one* introduces a literature review, aims and thesis organization.

Chapter two explains the theoretical part, neural networks and their basic classification types with focusing on the simple recurrent neural network SRN. The chapter also points out to the LabVIEW FPGA.

Chapter three presents the general discussion of the neural networks to simulate the behavior of the electronic circuits.

Chapter Four describes the Elvis II and its components and the design of ENN in LabVIEW FPGA that achieve the behavior of RC circuit, also the design of XOR circuit and their implementation on Elvis II. This chapter stands for the practical part of the work.

Chapter Five contains the conclusions that have been done through this thesis, and suggestions for the future works.

CHAPTER TWO

Neural Network Theory and Concepts

2.1 Introduction:-

An Artificial Neural network (ANN) is a collection of neurons in a particular arrangement or configuration. It is a parallel processor that can compute or estimate any function. Basically in an ANN, knowledge is stored in memory as experience and is available for use at a future time [5].

This chapter includes an introduction to both neural network, LabVIEW and FPGA. Section two discusses the classifications of artificial neural networks, while the concept of the recurrent neural network is introduced in section three, and finally, the principles of using LabVIEW for FPGA designs specially the design of neural networks can be found in section four.

2.2 Artificial Neural Networks:-

Many mathematical models for the (human) brain have been developed. The basic unit of a neural network is the processing unit. Fig (2-1) shows a brief description of the basic elements of the processing unit [12].

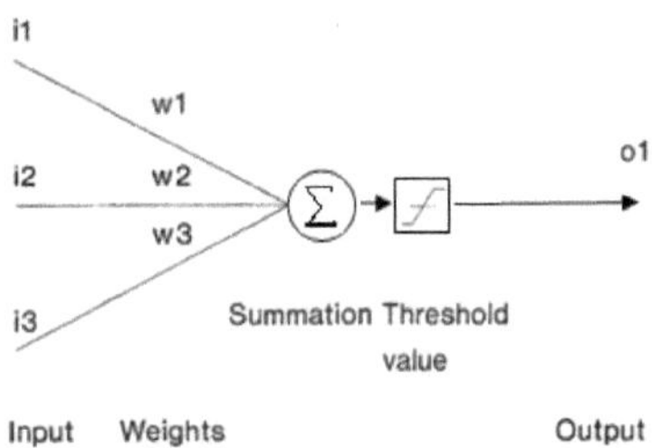

Figure (2-1): The Components of a Processing Unit [12]

An artificial neural network is an information-processing system that has certain performance characteristics in common with biological neural networks. Artificial neural networks have been developed as generalizations of mathematical models of human cognition or neural biology, based on the assumptions that:-

1- Information processing occurs at many simple elements called neurons.

2- Signals are passed between neurons over connection links.

3- Each connection link has an associated weight, which, in a typical neural net, multiplies the signal transmitted.

A neural network is characterized by (1) its pattern of connections between the neurons (called its architecture), (2) its method of determining the weight on the connections (called its training, or learning algorithm), and (3) its activation function [13].

The use of ANN offers the following advantages:

- **Non-linearity**: Since neurons are non-linear elements, the neural networks, which are made of interconnected neurons, are also non-linear. This characteristic is special in that it is distributed over the entire network.

- **Associations between the input and the output:** The networks are able to learn complex associations between input and output by means of a training process that is either supervised or non-supervised.

- **Adaptability:** The ability to adapt the synaptic weights of the network to change in the environment. A network that is trained to operate in a specific environment can easily be retrained when the conditions of that environment change.

- **Error tolerance:** A neural network is error-tolerant because the knowledge is divided over all the weights.

- **Neurobiological analogy:** The design of a neural network is motivated by the analogy with the human brain, which demonstrates that parallel processing is possible, fast, and powerful [14].

A neural network consists of large number of simple processing elements called neurons, units, cells or nodes. Each neuron is connected to other neurons by weights representing information being used by the net to solve a problem. Neural network can be applied to a wide verity of problems, such as storing and recalling data or patterns, classifying patterns, performing general mapping from input patterns to output patterns, grouping similar patterns, or finding solution to constrained optimization problem.

Each neuron has an internal state, called its activation or activity level which is a function of the inputs it has received. Typically, a neuron sends its activation as a signal to several other neurons. It is important to note that a neuron can send only one signal at a time, although that signal is broadcast to several other neurons [13].

Several key features of the processing elements of artificial neural networks are suggested by the properties of biological neurons:-

1- The processing element receives many signals.

2- Signals may be modified by weight at the receiving synaptic.

3- The processing element sums the weighted inputs.

4- Under appropriated circumstances (sufficient input), the neuron transmits a single output.

5- The output from a particular neuron may go to many other neuron (the axon branches).

Other features of artificial neural networks that are suggested by biological neurons are:

6- Information processing is local (although other means of transmission, such as the action of hormones, i.e of overall process control).

7- Memory is distributed :

    a- Long-term memory resides in the neurons' synapses or weights.

    b- Short-term memory corresponds to the signals sent by the neurons.

8- A synapses' strength may be modified by experience.

9- Neurotransmitters for synapses may be excitatory inhibitory.

Yet another important characteristic that artificial neural networks share with biological neural system is *fault tolerance*. Biological neural system is fault tolerant in two respects. **First**, we are able to recognize many input signals that are somewhat different from any signal we have seen before. An example of this is our ability to recognize a person in a picture we have not seen before or to recognize a person after a long period of time [13].

Second, we are able to tolerate a damage to the neural system itself. Humans are born with as many as 100 billion neurons. Most of these are in the brain, and most are not replaced when they die. In spite of our continuous loss of neurons, we continue to learn. Even in case of traumatic neural loss, other neurons can sometimes be taken over the functions of the damaged cells. In a similar manner, artificial neural networks can be designed to be investigative to a small damage to the network, and the network can be retrained in case of significant damage (e.g., loss of data and some connection) [13].

2.3 <u>Classifications of Neural Network</u>:-

There are various architectures of artificial neural networks. This depends on the number of layers, connection pattern and learning algorithm. Fig (2-2) shows the classification of neural networks.

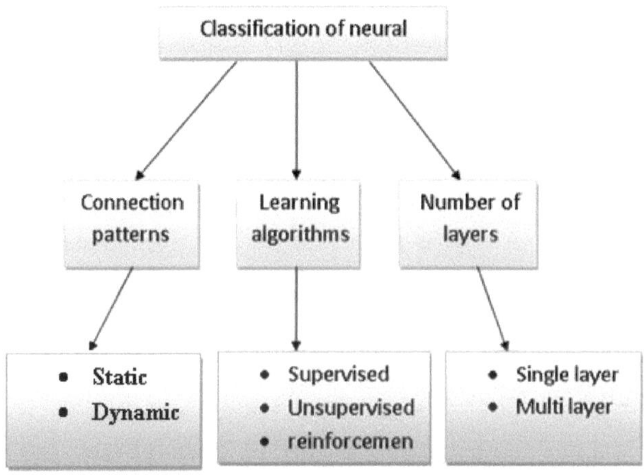

Figure (2-2): Neural Network Classification

The network architecture or topology comprises number of nodes in hidden layers, network connections, initial weight assignments, and activation functions: it plays a very important role in the performance of the ANN, and usually depends on the problem at hand . Fig (2-3) shows a simple ANN and its constituents. In most cases, setting the correct topology is a heuristic model selection, whereas the number of input and output layer nodes is generally suggested by the dimensions of the input and the output spaces, determining the network complexity is yet again very important. Too many parameters lead to poor generalization (over fitting), and too few parameters result in inadequate learning (under fitting) [11].

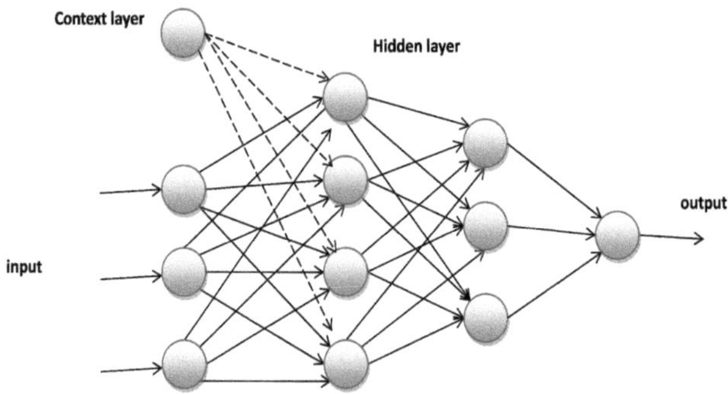

Figure (2-3): Simple Neural Network [8].

Input Layer: Size depends on Problem Dimensionality.

Hidden Layer: A design parameter must decide number of layers, and size for each layer creates a non-linear generalized decision boundary.

Output Layer: Size depends on number of classification categories.

Bias: Further generalizes the decision boundary Net

Activation: It weighted sum of the input values at respective hidden nodes.

Activation Function: It decides how to categorize the input to a node into a possible node output incorporating the most suitable non-linearity.

Network Learning: Training an untrained network: Several training methods are available.

Stopping Criterion: Indicate when to stop the training process; e.g., when a threshold MSE is reached or maximum number of epochs used.

Every ANN consists of at least one hidden layer in addition to the input and the output layers. The number of hidden units governs the expressive power of the net and thus the complexity of the decision

28

boundary. For well-separated classes fewer units are required and for highly interspersed data more units are needed. The number of synaptic weights is based on the number of hidden units. It represents the degrees of freedom of the network. Hence, we should have fewer weights than the number of training points. As a rule of thumb, the number of hidden units is chosen as $\left[\dfrac{n}{10}\right]$, where n is the number of training points. But this may not always hold a true and a better tuning which might be required depending on the problem [15].

Tan-sigmoid Transfer Function:-

Basically the transfer function of a neuron performs two very important tasks: first, it restricts the output of the neuron to avoid its values from turning too high, and the second is that it provides characteristics of non-linearity which is very important for NN [16]. In this thesis, the tangent sigmoid activation function has been used. This function is shown in Fig (2-4) [17].

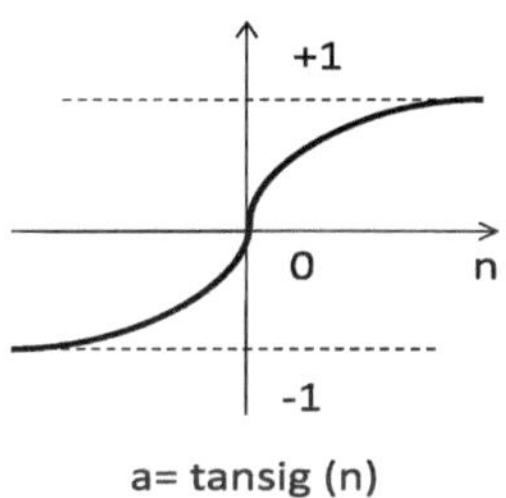

Figure (2-4): Tan-sigmoid Transfer Function [17]

This function takes the input (which may have any value between plus and minus infinity) and the output value into the range - 1 to 1, according to the expression[17]:

$$a = \frac{e^n - e^{-n}}{e^n + e^{-n}} \qquad \ldots\ldots\ldots\ldots (2.1)$$

The algorithm of tan-sigmoid is as in the following equation [17]

$$a = tansig(n) = 2/(1 + \exp(-2 * n)) - 1 \quad ..... \quad (2.2)$$

The tan-sigmoid activation function is commonly used in multilayer neural networks that are trained by the back-propagation algorithm, since this function is differential. The tan-sigmoid function is not easily implemented in digital hardware because it consists of an infinite exponential series. Many researchers used a lookup table to implement the tan-sigmoid function. The drawback of using lookup table is the great amount of hardware resources needed. A simple second order nonlinear function presented by kwan [18], can be used as an approximation to a sigmoid function. This nonlinear function can be implemented directly using digital techniques. The following equation is a piecewise nonlinear function which has a tan-sigmoid transition between the upper and lower saturation regions [19]

$$f(n) = \begin{cases} n\,(B - g.n) & for\ 0 \leq n \leq L \\ n\,(B + g.n) & for\ -L \leq n < 0 \end{cases} \quad .... \quad (2.3)$$

where **B** and **g** represent the slope , and the gain of the nonlinear function *f(n)* between the saturation regions -L and L [18]. The hardware implementation of f (n) can be carried out according to the signal-flow graph as shown in Fig (2-5)

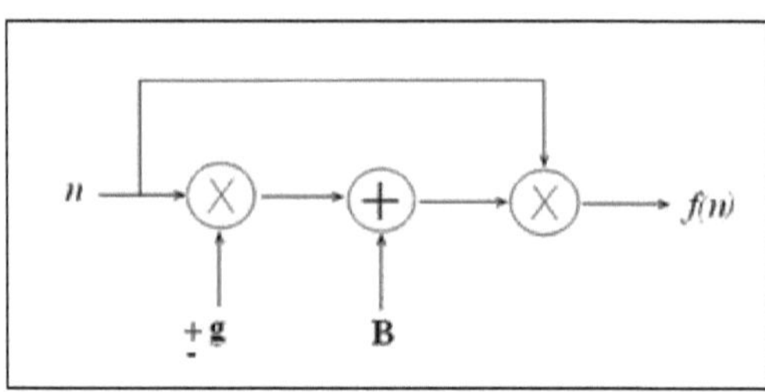

Figure (2-5): signal-flow graph of f(n) [18]

The design of tan-sigmoid in LabVIEW FPGA will be seen in chapter four.

2.3.1 Number of Layer

NNs can be classified according to number of layer into single layer and multi-layer neural network. In single layer, the network consists of an input layer of sources neuron, and an output layer of computational neurons as shown in Fig (2-6a). On the other hand, in Fig (2-6b), in addition to the input and output layer, the network have at least one middle or hidden layer of computational neurons .This is called multi-layer network [13].

Neural networks with hidden units are universal approximators, which mean that, in theory, they are capable of learning an arbitrarily accurate approximation to any unknown function, provided that they increase in complexity at a rate approximately proportional to the size of the training data [20].

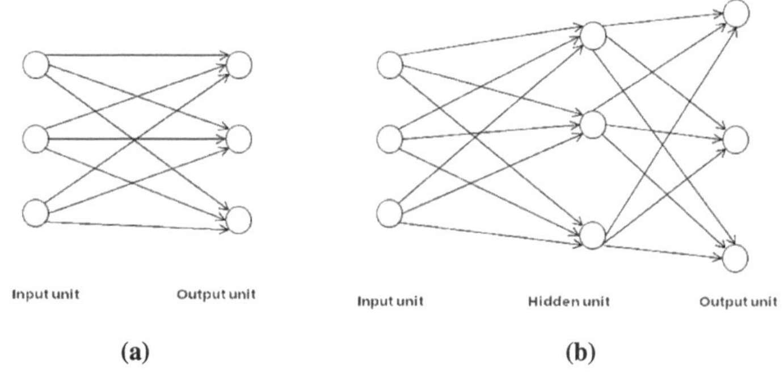

Figure (2-6): (a) Single- Layer Neural net, (b) Multilayer Neural net [20]

2.3.2 Network in Connection Pattern

Neural networks can be classified into dynamic and static categories. Static (feed-forward) networks have no feedback elements and contain no delays; the output is calculated directly from the input through feed-forward connections. In dynamic networks, the output depends not only on the current input to the network, but also on the current or previous inputs, outputs, or states of the network [4].

Networks can be subdivided into feed-forward networks Fig(2-7a) where data flow one way from input to output, and recurrent networks for which the output values are fed back to input Fig (2-7b). Some networks Fig (2-7c) allow connections between neurons in the same layer (lateral connections) [21].

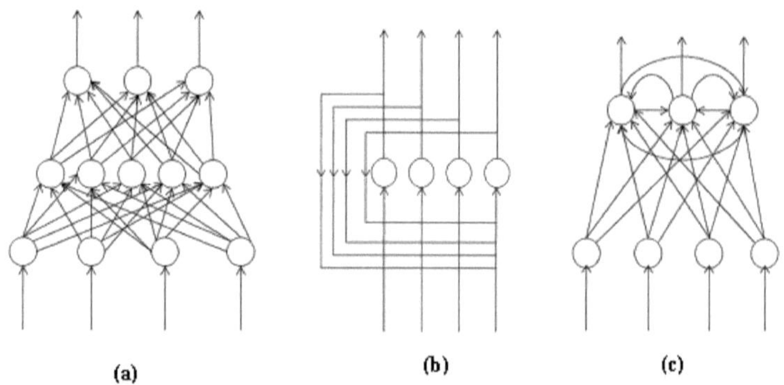

Figure (2-7): Three Neural Networks with Different Types of Connections: (a) A Feed-Forward Network, (b) A Recurrent Network, and (c) A Network with Lateral Connections [21].

In the next section, more details about connection patterns in learning algorithm will be seen.

2.3.3 Network in Learning Algorithms

There are three major frameworks of learning: *supervised learning,* based on the output error signal, *reinforcement learning,* based on the scalar reward signal, and *unsupervised learning,* based on the statistical features of the input signal [22]. Our main focus will be on the supervised learning algorithms for recurrent networks.

The main goal of supervised learning is to minimize the error function:

$$E = \frac{1}{2} (yd(k) - ya(k))^2 \;\dots\dots\; (2.4)$$

where yd(k) is the target (desired output) and ya(k) is the actual output of the FNN. Also, in supervised learning, neural network weights are tuned to make the outputs as close to those of the identified process when responding to the same inputs. When all weights converge to an equilibrium state at which the neural model approximates the actual process's behavior, the neural network is said to be globally stable. If the weights converge to incorrect values, the neural network model is considered to converge to a local minimum and may not represent the actual system. A neural network is said to be unstable if the weights diverge during the training process. A properly designed learning rule should guarantee the global convergence of the network learning process [23, 24].

The fundamental difference between a system that learns and one that merely memorizes is that the learning system generalizes to unseen examples. Much of our concern is with supervised learning, getting a network to behave in a way that successfully approximates some specified pattern of behavior or input-output relationship. In particular, much emphasis has been placed on feed-forward networks which have no loops, so that the output of the network can depend on its input alone, since there is then no internal state defined by reverberating activity. Rather the "input" is the initial state of the network, and the "output" is the "attractor" or equilibrium state to which the network then settles. For neuron whose output is a sigmoid function of the line is combination of their inputs, the memory capacity of the associative memory is approximately 0.15n,where n is the number of neurons in the net.

Historically, the earliest forms of supervised learning involved changing synaptic weights to oppose the error in a neuron with a binary output (the perceptron error correction rule),or to minimize the sum of squares of errors of output neurons in a network with real-valued outputs [22].

2.4 <u>Static Neural Network:-</u>

Static (Feed-forward) neural network, all data flows in a network in which no cycles are present (have no feed-back elements and contain no delays); the output is calculated directly from the input through feed-forward connections and it means that the neural network behavior depends only on the current input network. (Perceptron, Redial Basic Function, Back-propagation, etc.) are examples of the feed-forward neural network. The training of static networks will be discussed in the following section "Back-propagation NN". Fig (2-8) shows feed-forward neural network connection patterns [17].

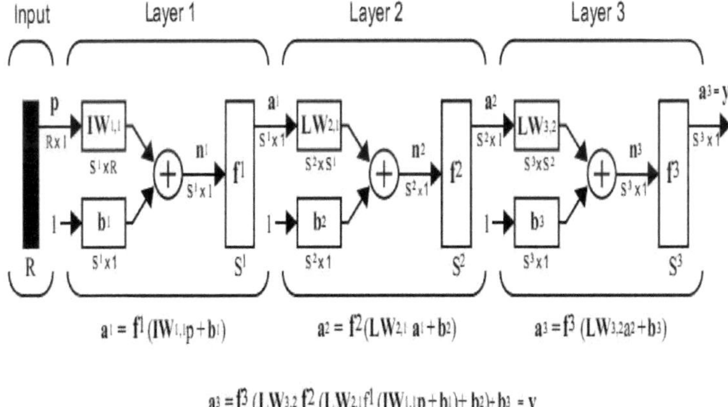

$$a^1 = f^1(IW_{1,1}p + b^1)$$

$$a^2 = f^2(LW_{2,1}a^1 + b^2)$$

$$a^3 = f^3(LW_{3,2}a^2 + b^3)$$

$$a^3 = f^3(LW_{3,2}f^2(LW_{2,1}f^1(IW_{1,1}p + b^1) + b^2) + b^3) = y$$

Figure (2-8): Architecture of Feed-Forward Neural Network [17].

Back-propagation Neural Network

It is a training method that is called back-propagation of errors or the generalized delta rule; it is simply a gradient descent method to minimize the total squared error of the output computed by the network. As in the case with most neural networks, the aim is to train the network to achieve a balance between the ability to respond correctly to the input patterns that are used for training (memorization) and the ability to give reasonable (good) responses to input that is similar , but not identical , to that used in training generalization. The training of the network by back-propagation involves three stages: the feed-forward of the input training patterns , the calculation and back-propagation of the associated weights, and the adjustment of these weights. After training, application of the network involves only the computations of the feed-forward phase [13].

The algorithm of back-propagation is as follow:

Step0. Initialize weights.

Step1. While stopping condition is false, do Steps 2-9.

Step2. For each training pair, do Steps 3-8.

Feed-forward:

Step3. Each input unit (Xi, i=1, ….. n) receives input signals x_i and
broadcasts these signal to all units in the layer above (the
hidden units).

Step4. Each hidden unit (Z_i , i=1, …..p) sums its weighted input
Signals,

$$z_{-inj} = v_{oj} + \sum_{i=1}^{n} x_i\, v_{ij} \;\text{…….}\; (2.5)$$

Applies its activation function to compute its output signal,

$$z_j = f(z_{-inj}) \;\text{…….}\; (2.6)$$

And sends this signal to all units in the layer above (output
units).

Step5. Each output units (Yk , k=1, ….., m) sums its weighted input
Signals,

$$y_{-ink} = w_{ok} + \sum_{i=1}^{n} zj\, w_{jk} \;\text{…….}\; (2.7)$$

And applies its activation function to compute its output signal

$$y_k = f(y_{-ink}) \;\text{…….}\; (2.8)$$

Back-propagation of error:

Step6. Each output unit (Yk , k=1, ….., m) receives a target pattern
corresponding to the input training pattern, computes its error

information terms

$$\delta_k = (t_k - y_k)f'(y_{\_ink}) \quad \ldots\ldots (2.9)$$

Calculates its bias correction term (used to update w_{Ok} later),

$$\Delta w_{jk} = \alpha\delta_k z_k \quad \ldots\ldots (2.10)$$

And sends δ_k to units in the layer below.

Step7. Each hidden unit (Zj, k=1, ..., p) sums its delta inputs (from units in the layer above) ,

$$\delta_{\_inj} = \sum_{k=1}^{m} \delta_k w_{jk} \quad \ldots\ldots (2.11)$$

Multiplies by the derivative of its activation function to calculate its error information term,

$$\delta_j = \delta_{\_inj}f'(z_{\_inj}) \quad \ldots\ldots (2.12)$$

Calculate its weight correction term (used to update v_{ij} later),

$$\Delta v_{ij} = \alpha\delta_j x_i \quad \ldots\ldots (2.13)$$

And calculates its bias correction term (used to update v_{oj} later

$$\Delta v_{oj} = \alpha\delta_j \quad \ldots\ldots (2.14)$$

Update weights and biases:

Step 8. Each output unit (Yk , k=1, ..., m) updates its bias and weights (j=0,.......,p):

$$w_{jk}(new) = w_{jk}(old) + \Delta w_{jk} \quad \ldots\ldots (2.15)$$

Each hidden unit (Zj , j=1, ..., p) updates its bias and weights (i=0,........,n):

$$v_{ij}(new) = v_{ij}(old) + \Delta v_{ij} \quad \ldots\ldots (2.16)$$

Step9. Test stopping condition.

Note that in implementing this algorithm, separated arrays should be used for the deltas for the output units (step 6, δ_k) and the deltas for the hidden units (step 7,δ_j).

2.5 <u>Dynamic Neural Network:-</u>

Dynamic networks can also be divided into two categories: those that have only Feed-forward connections and those that have feed-back, or recurrent, connections. The recurrent dynamic networks typically have a longer response than the feed-forward dynamic networks. Dynamic networks are generally more powerful than static networks (although somewhat more difficult to train). Because dynamic networks have memory, they can be trained to learn sequential or time-varying patterns [13,17].

2.5.1 Recurrent Neural Network:-

Recurrent Neural Networks (RNNs) are nonlinear or linear dynamic systems. They can be simulated in software on computers or implemented in hardware (analog or digital). RNNs have a cycle that can be connected from output of hidden unit to the input this depends on the type of the network. Recurrent neural network has feed-back connections with delays and it is also called dynamic neural network which means that the neural network behavior depends not only on the current input (as in feed-forward networks) but also on the previous operations of the network. Some properties of recurrent neural networks are investigated, especially their ability to perform classification of sequences of data [4]. (Hopfield, Boltzman, Elman, etc.) are examples of recurrent neural network.

A neural network with acyclic topology consists of no feed-back loops. Such an acyclic neural network is often used to approximate an on linear mapping between its inputs and outputs. A neural network with cyclic topology contains at least one cycle formed by directed arcs. Due to the feed-back loop, a recurrent network leads to a nonlinear dynamic system model that contains internal memory. Recurrent neural networks often exhibit complex behaviors and remain an active research topic in the field of artificial neural networks [20]. Figure (2-9) shows the architecture of recurrent neural network.

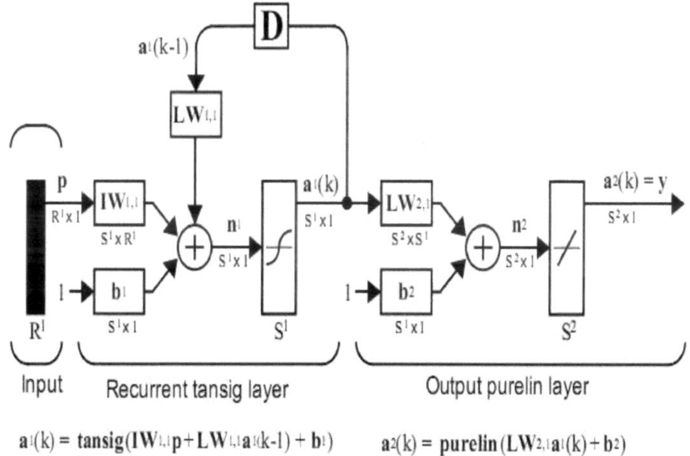

Figure (2-9): Architecture of Recurrent Neural Network [17].

In the case of working with RNN, the system will have a series of additional parameters to the Feed-forward neural network. In total, it will have the following parameters:

• **Percentage of new patterns**. How many new patterns will be used apart from the training patterns. In the case of not working with time series, these new patterns are checked so that they shall be different from the training patterns. If working with a time series, these new random patterns are added at the end of the series used for training.

• **Probability of change**. The process for creating new test cases focuses on generating new sequences from the training data. A sequence of inputs is chosen at random from the training file. For each input, the same value is maintained or a new one is generated at random, following a certain probability.

• **Minimum error for survival**. Once the process has started to extract rules, and as these are generated, they reflect the patterns with which they are being trained. Once the rules extracted sufficiently reflect the behavior of the ANN with a given pattern, this pattern is no longer necessary, and so it is discarded and a new one created. This parameter indicates when a pattern should be discarded, when the error value obtained for the rules in this specific pattern is below a certain threshold.

• **Number of feedback outputs**. When working with recurrent networks, the aim is also studying the behavior of the network by taking the outputs returning and using them as inputs. This parameter indicates the number of patterns (the group of patterns whose size is indicated by the parameter "percentage of new patterns") that will not be generated at random, but which instead will be the result of simulating the network, taking each input as the output in the previous moment [14].

Types of tasks for which RNNs can, in principle, be used:

❖ System identification and inverse system identification.
❖ Filtering and prediction.

❖ Pattern classification.

❖ Stochastic sequence modeling.

❖ Associative memory.

❖ Data compression.

Some relevant application areas:

❖ Telecommunication.

❖ Control of chemical plants.

❖ Control of engines and generators.

❖ Fault monitoring, biomedical diagnostics and monitoring.

❖ Speech recognition.

❖ Robotics, toys and edutainment.

❖ Video data analysis.

❖ Man-machine interfaces [25].

2.5.1.1 Fully Recurrent Neural Network (FRNN)

The network is called Fully Recurrent if the output of all neurons is recurrently connected to all neurons in the network [20]. Another name for this type of network is the Real-Time Recurrent Network Which is a gradient- descent method which computes the exact error gradient at every time step [25] .

2.5.1.2 Partially Recurrent Neural Network (Simple Recurrent Network (SRN))

The network described in the previous sections is limited to realizing static mappings. Once a network is configured it only maps the input at time T on the output according to some learned mapping. Hence, the network is not capable of taking into account inputs that it processed earlier unless they were presented during the learning period. One way to eliminate this restriction is to use a shift register that consists of N buffers, each capable of storing a single value. Each time a new input sample arrives the buffers are shifted, i.e., buffer N becomes N-I, N-1 becomes N-2 etcetera. The contents of buffer N are forgotten and the new sample is stored in buffer 1. A network with N buffers as input neurons is capable of processing temporal information restricted to the last N samples. First of all this solution is very awkward since it requires shift registers that are physically limited, meaning that only a limited number of samples can be retained. Secondly, this solution introduces a translation problem since a pattern can begin at N different positions [12]. Another way to eliminate this restriction without using buffers is to use outputs at time t-1 as input at time t. These may either be outputs from neurons in the output layer but can just as well be taken from any other neuron in the network. Such a network is called a recurrent network and since its structure has remained the same it can still be trained using the Back Propagation procedure. One particular kind is called Simple Recurrent Network (SRN) and has been studied by Elman (1988). This network contains an input layer, a hidden layer and an output layer. The input layer is divided into input neurons that actually serve as the network input and so-called context neurons that are connected to the hidden neurons. For each hidden neuron there is exactly one context neuron and after each iteration the output of a hidden neuron is copied to the output of its corresponding context neuron. The structure of this recurrent network is depicted in Fig (2-10) [12].

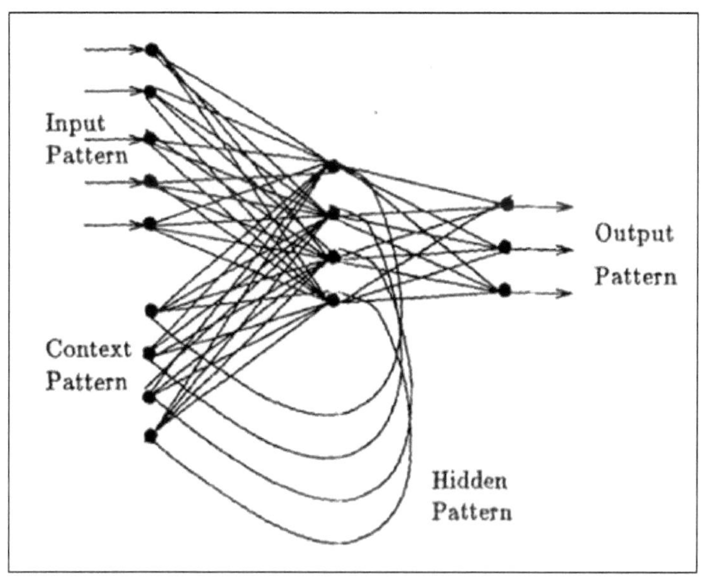

Figure (2-10): A Simple Recurrent Network (Elman) [12].

Several neural networks have been developed to learn sequential or time-varying patterns. Unlike the recurrent nets with symmetric weights or the feed-forward nets, these nets do not necessarily produce a steady- state output [13]. In this work, Elman neural network will be used to simulate the behavior of electronic circuit. It can be simple recurrent network (SRN) or partially recurrent neural network. The weights from the context units to the hidden units are trained exactly by the same manner as the weights from the input units to the hidden units. Thus, at any time step, the training algorithm is the same as for standard back-propagation [13].

RNNS can generate and store temporal information and sequential signals. A commonly used recurrent architecture is the Elman model [21]. Fig (2-11) shows the layers of Elman, which acts as a short-term memory in the system.

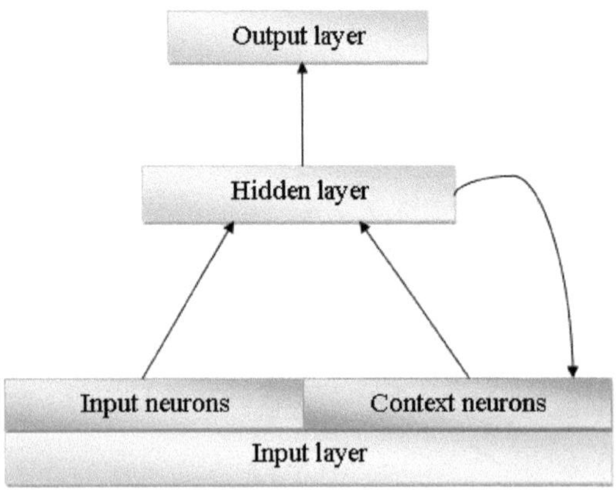

Figure (2-11): Architecture of Elman neural network

Algorithm for training Elman neural network is the back-propagation algorithm used in computing the error. Arthritically, there is a feedback connection from the hidden layer to the context layer (which act as short term memory) and then this fed to the hidden layer at$(t + 1)$.

Figure (2-12) shows that a long ENN where by a back propagation is used to calculate the derivatives of the error (at each output unit) by unrolling the network to the beginning. At the next time step t+1 input represented the context units contain values which are exactly the hidden unit values at time t (and the time t-1, t-2 …) and these context units provide the network with memory [6]. Therefore, the ENN network is converted into a dynamical network that is efficient in the use of temporal information of the input sequence, both for classification as well as for prediction [26].

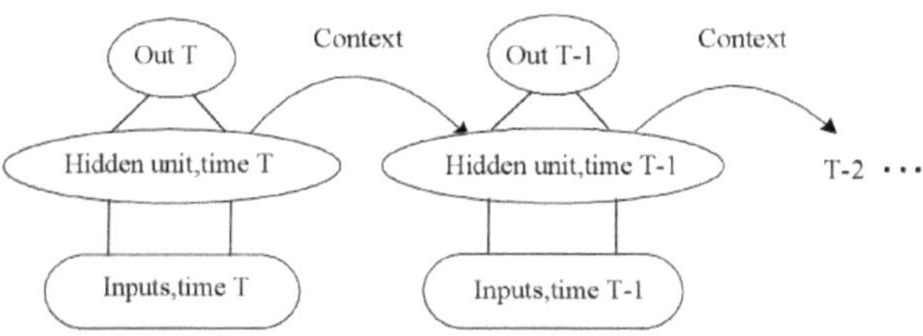

Figure (2-12): Unroll the ENN through Time [27].

2.6 <u>Field Programmable Gate Array (FPGA)</u>:-

ANNs are biologically inspired and they require parallel computations in their nature. Microprocessors and DSPs are not suitable for parallel designs. Designing fully parallel modules can be available by ASICs and VLSIs but it is expensive and time consuming to develop such chips. In addition ,the design results in an ANN suited only for one target application. FPGAs not only offer parallelism but also flexible designs, savings in cost and design cycle [28].

NI Digital Electronics FPGA Board can be programmed with Xilinx ISE tools and the NI LabVIEW graphical system design environment. Xilinx ISE tools work with both the Verilog hardware descriptive language and VHDL. Because LabVIEW is based on the graphical, data flow paradigm, students can quickly design and prototype their systems on the FPGA without having to learn new languages such as Verilog or VHDL. LabVIEW and LabVIEW FPGA feature full support for the I/O on the board, making it easy to include any I/O on the board that students may want in their designs with a drag-and-drop interface. LabVIEW also provides a one-click compile, synthesize, place, and route design [27].

In this thesis, the designer use LabVIEW to design neural network.

CHAPTER THREE

Neural Network Design and Implementation of Electronic Circuit

3.1 Introduction:-

Since recurrent neural network is used to simulate time series signals, it can be used to simulate the behaviour of electronic circuits. The typical electronic circuit that very well demonstrates the behaviour of time series signals is the RC circuit.

This chapter introduces the simulation of recurrent neural network. The following articles explain the behaviour of signals through neural network. The last sections contain the proposed design and results. Results are given using **MATLAB® R2010a** environment.

Signal processing is used to analyze the signals in discrete and continuous time. The Analog signal processor involves linear and non linear electronic circuit such as passive filter, active filter, adaptive mixers integrator, differentiator, delay lines, amplifiers, voltage_ controlles, voltage _ controlled filters and so on [34]. Here, neural network is used for processing analog signals. The neural network system is used for RC circuit behavior analysis.

3.2 Simple Memory Restoration of Patterns:-

Neural network must contain memory in order process temporal information. There are two basic ways to build memory into the neural network. The first one is to introduce time delays in the network and to adjust their parameters during the learning phase. The second way is to use positive feed-back, i.e making the network recurrent [35].

Associative memory networks come in a variety of models. The most important classes of associative memories are static and dynamic memories. Table (3-1) shows a comparison between the static and dynamic memory in neural network [36].

Table (3-1): Comparison between static and dynamic memory [36]

Static memory	Dynamic memory
recall an output response after an input has been applied in one feed-forward pass	produce recall as a result of output/input feed-back interaction, which requires time
Theoretically, without delay	with delay
implement a feed-forward operation of mapping without a feed-back	implement a feed-forward operation of mapping with a feed-back
called non-recurrent	Called recurrent
performs the mapping as in Equation $$\mathbf{V^t = M_1\,[X^t]}$$	performs the mapping as in Equation $$\mathbf{V^{t+1} = M_1\,[X^t,\,V^t]}$$
Fig (3-1a) show block diagrams for static memory	Fig (3-1b) show block diagrams for dynamic memory

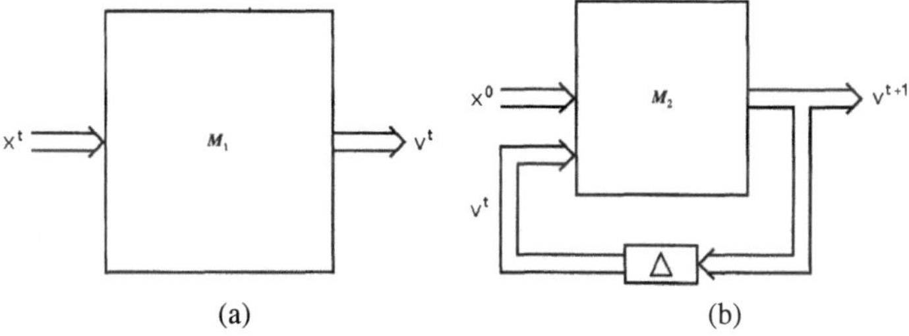

Figure (3-1): Block diagram representation of associative memories: (a) feed-forward network, (b) recurrent auto associative network [36].

To characterize memories in different architectures, two dimensions, *depth* and *resolution* has been considered. Roughly, *depth* refers to how far into the past the memory stores information relative to the memory size, and *resolution* how accurately information concerning the individual elements of the input sequence is preserved. In time series prediction with neural networks the main problems have decided prediction order and the structure of the network. These problems remain mostly the same for the architectures studied. In addition, the stability of the model and the learning algorithm must be considered [34].

3.3 <u>The Proposed Design And Training of The Network:-</u>

In this thesis, RNN, which uses positive feed-back as a memory, has been used.

Recurrent neural networks (RNNs) have been efficient identification tool in many areas since they have dynamic memories. A RNN with a dynamic memory is more suitable for representing a dynamic system, which has a dynamic mapping between its outputs and inputs. RNNs generally require less neurons in the neural structure and less computation time. Moreover they have a low probability of being affected by external noise. Because of these features, RNNs have attracted the attention of researchers in the field of dynamic system identification [6].

In Elman network, positive is feedback that used to construct memory in the network is fed from the output of hidden layer to the input. The network that is proposed to represent the behaviour of RC circuit consists of two recurrent layers and an output layer. Also, units called context units that work as short term memory to save previous output values of hidden layer neurons. Context unit values are then fed back fully connected to hidden layer neurons and thus they serve as additional inputs to the network [35]. The output of context layer at time (T) equals to the output of hidden layer at time (T-1). Networks output layer values are not fed back to the network. The layers of the network and the designed neural network can be seen in Fig (3-2) and Fig (3-3) respectively.

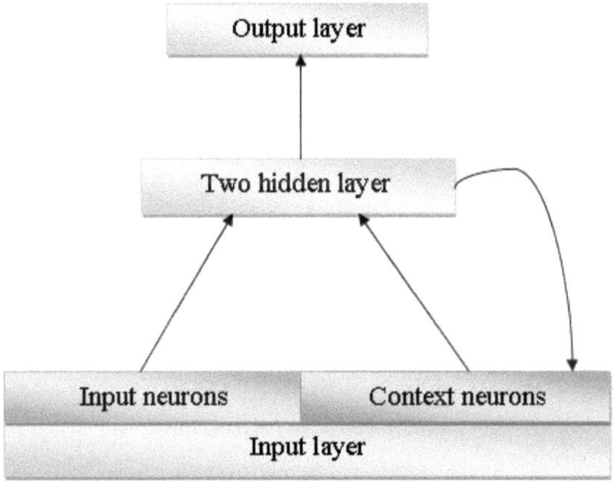

Figure (3-2): The layers of Elman neural network

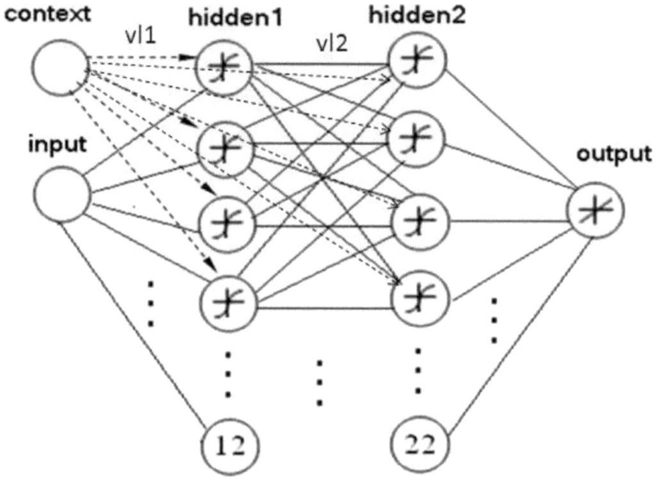

Figure (3-3): The Designed Neural Network

From Fig (3-3), it is clear that the first recurrent layer consists of (12) neurons and the second recurrent layer has (22) neurons. Output layer is Linear, while the hidden layer uses the tan-sigmoid function. The vectors WL1,WL2,Wo contain the weights of feed-forward connection while VL1,VL2 contain the weights of the feedback connections.

The Elman network has a high depth and low resolution memory, since the context units keep exponentially decreasing trace of past hidden neuron output values [35].

The algorithm for training the network has the following steps:-
Step1. Feed the input vector **input** .
Step2. Initialize randomized weight matrix **wl1,wl2,wo vl1**and **vl2** and bias **b1, b2**and **b3**.
Step3. Compute the vectors **z1=wl1*input**.
Step4. Feed-back **z1** to the input of first recurrent layer.

Step5. Compute **zi1=z1*VL1** and pass zi1 into n tan-sigmoid

      functions.

Step6. Compute **z2=WL2*zi1**.

Step7. Feed-back **z2** to the input of second recurrent layer.

Step8. Compute **zi2=z2*VL2** and pass **zi2** into n tan-sigmoid

      functions.

Step9. Calculate $out = zi * wo$.

Step10. Calculate the error by the steps of back-propagation of error

      mentioned in chapter two.

Step11. Repeat the neural network training until the error

      approximately equal to 0.

Many trials were applied to different types of networks before the designer discover the correct structure of the network that show the expected results. Firstly, back-propagation NN had been used to design the computation electronic circuit. After training, different waves (sin, square, triangular waves) were fed to the network but the results were not as expected. Therefore, this type of network was rejected.

The set of outcomes of any electronic circuit or device simulations can be used as the target data set for a neural network. Here, the outcomes of RC circuit will be used as a target data for the neural network. The electronic circuit that has been considered is shown in Fig (3-4):

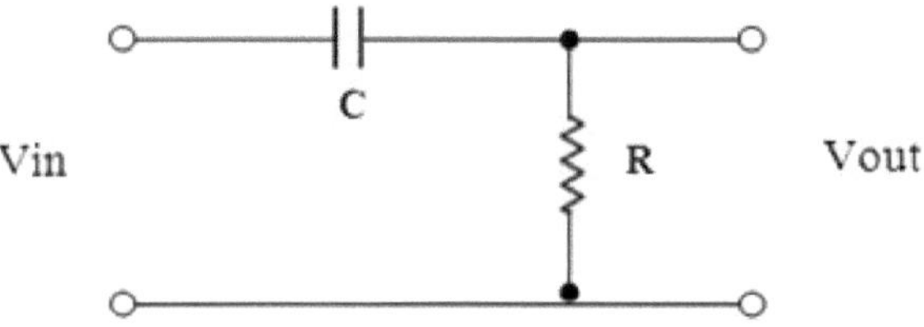

Figure (3-4): Elementary Connection Circuit

The differential equations relating Vout and Vin, the output and the input voltage, respectively, which describes the behaviour of our RC circuit, is

$$Vout = iR \qquad\qquad \text{........ (3.1)}$$

$$Vout = RC\ \frac{d(Vin-Vout)}{dt} \qquad\qquad \text{........ (3.2)}$$

To explore the input/output values for different input waveforms, Eq. (3.1) was built by Simulink (MATLAB simulation program) in order to use its input/output data to train the RNN. The simulation model of equation (3.2) as shown in Fig (3-5) consists mainly of differentiator and amplifier of gain=RC.

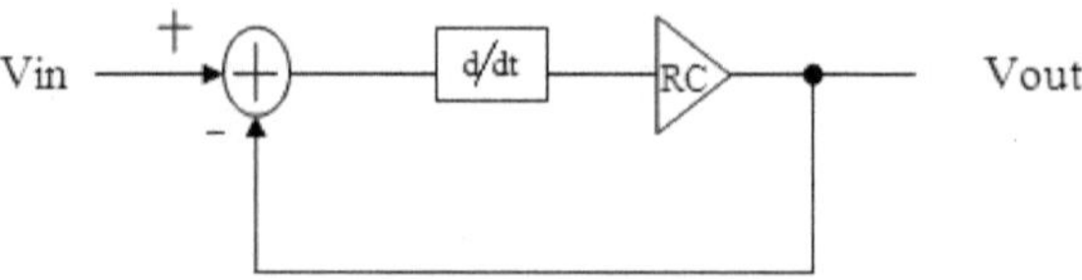

Figure (3-5): RC Circuit in Simulink

Figure (3-6) shows the structures of the designed Elman NN that consists of three-layers (2-recurrent layers and one output layer).

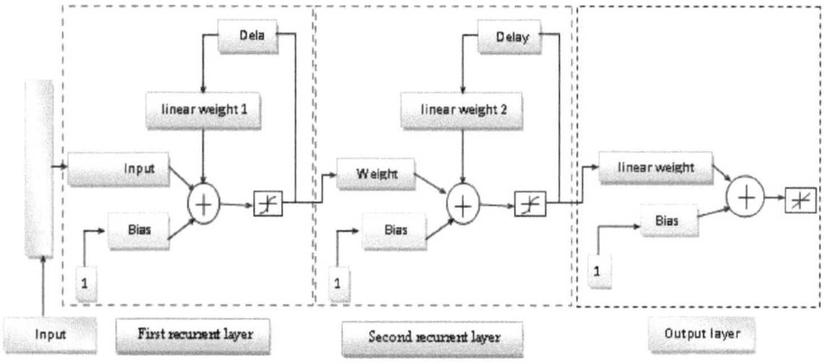

Figure (3-6): Three-layer Elman network

3.4 Results of Neural Network Training:-

In this section, all expected results of the designed neural network after training are shown and explained. Also, the waveforms of the input and the output of the RC circuit, which represent the training set and the result of the neural network (the waveforms before training) are presented. And then after training, the neural network is tested by feeding different waveforms (sinusoidal, triangular and square waves), and finally, the network outputs are compared with the target outputs of the real RC circuit.

Firstly, Fig (3-7a) shows the input and the output of the RC circuit. It is clear that for sine function input, the output is a cosine function. Samples of these two waves with rates greater than Nyquist rate are trained by the RNN. Fig (3-7b) shows the input and the output waveforms of the RNN after training when the network is tested by sine function input. It is quite clear that the RNN acts as the RC circuit.

53

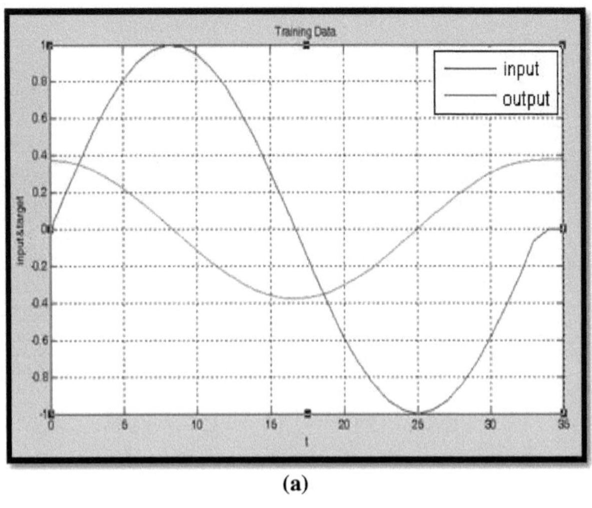

(a)

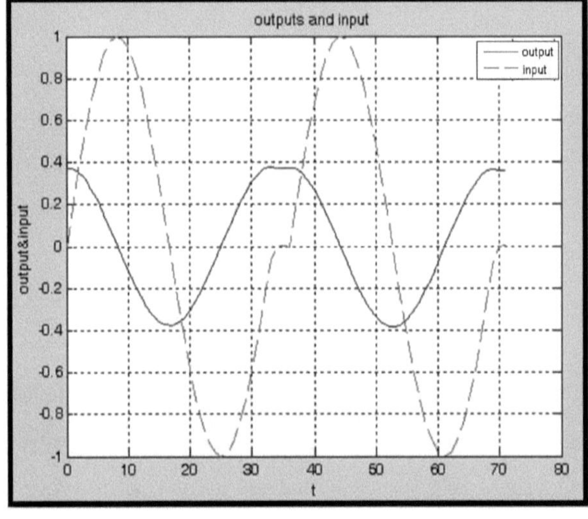

(b)

Figure (3-7): The input sine wave and the output is cosine wave

(a) Training signals (RC input/output) and (b) test signals (network

input/output)

The final design of the neural network of the sine wave has 5 neurons

in the first recurrent layer, 8 neurons in the second recurrent layer and one neuron in the output layer. By MATLAB, the training function is 'traingdx' and the performance function is 'mse'.

In Fig (3-8a) and Fig (3-9a), (square and triangular waveforms) represent the inputs and (impulse and square waveforms) represent the outputs of the RC circuit respectively. While Fig (3-8b) and Fig (3-9b) which are the same waveforms are coming out from the RNN after training.

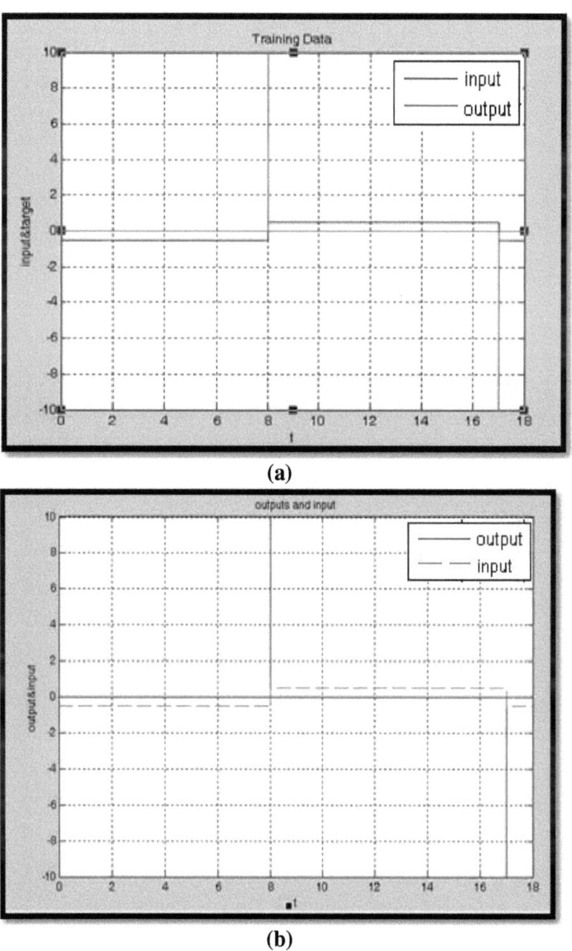

(a)

(b)

The final design of the neural network of the square wave has 5 neurons in the first recurrent layer, 7 neurons in the second recurrent layer and one neuron in the output layer. By MATLAB, the training function is 'traingdx' and the performance function is 'mse'.

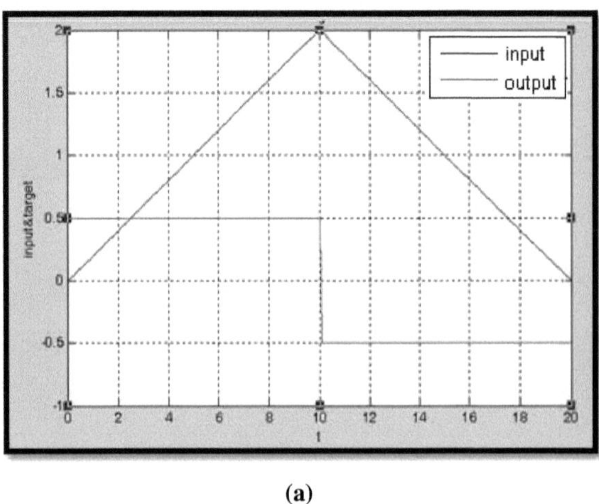

(a)

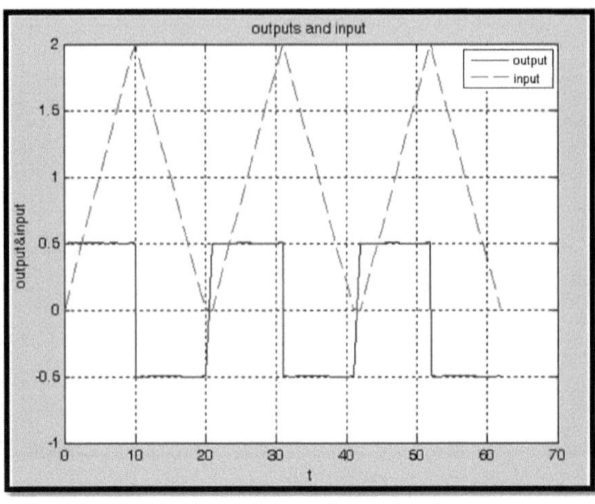

Figure (3-9): The input is triangular wave and the output is square wave (a) Training signals (RC input/output) and (b) test signals (network input/output)

The final design of the neural network of the triangular wave has 15 neurons in the first recurrent layer, 17 neurons in the second recurrent layer and one neuron in the output layer. By MATLAB, the training function is 'traingdx' and the performance function is 'mse'.

At the last, the final design of the RNN is trained in all types of signals (sine, triangular and square) all together. Figure (3-10) shows the training signals that are coming out from the RC circuit. It can be seen that the NN output converges towards the target as the number of training epochs increases for all signals. The performance goal was eventually met in epoch (69977). This proves that the error could be made smaller by more training.

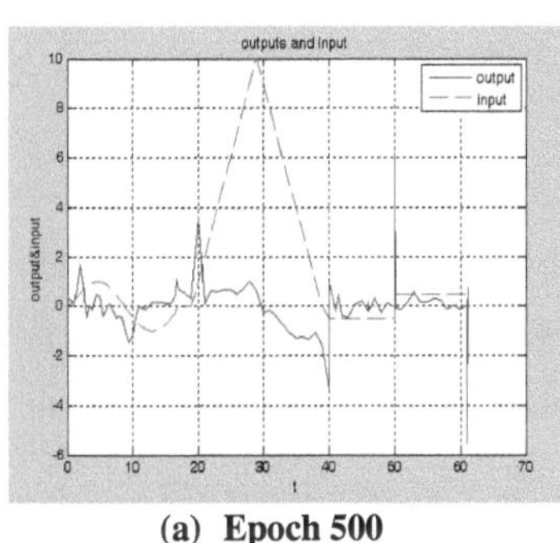

(a) **Epoch 500**

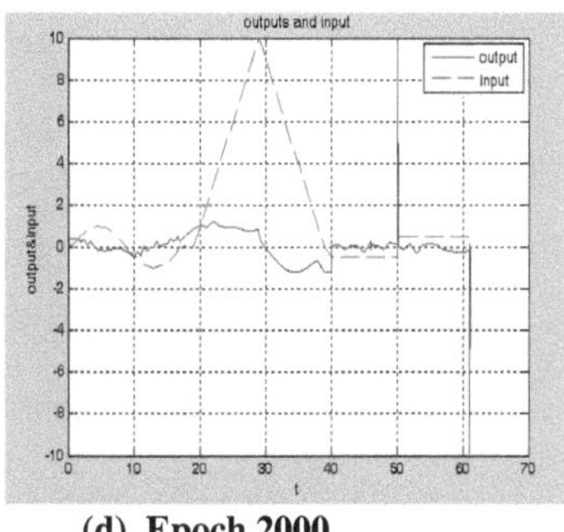

(d) **Epoch 2000**

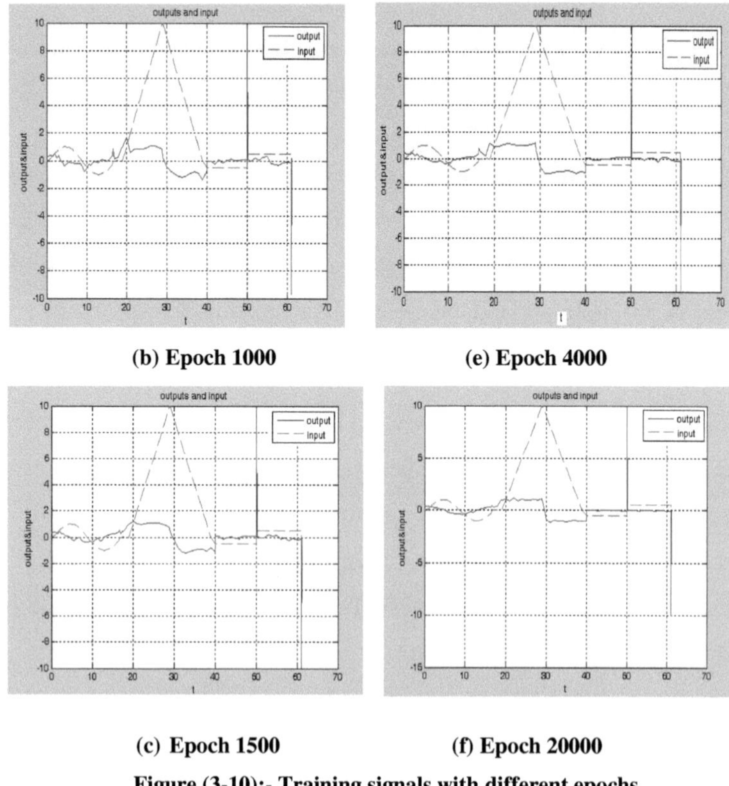

(b) Epoch 1000   **(e) Epoch 4000**

(c) Epoch 1500   **(f) Epoch 20000**

Figure (3-10):- Training signals with different epochs

Figure (3-11a) shows the training signals that are coming out from the RC circuit, and Fig (3-11b) shows the input/output of the RNN after training. It is clear that the RNN generates the same waveforms that are the output from the RC circuits when both have the same inputs.

58

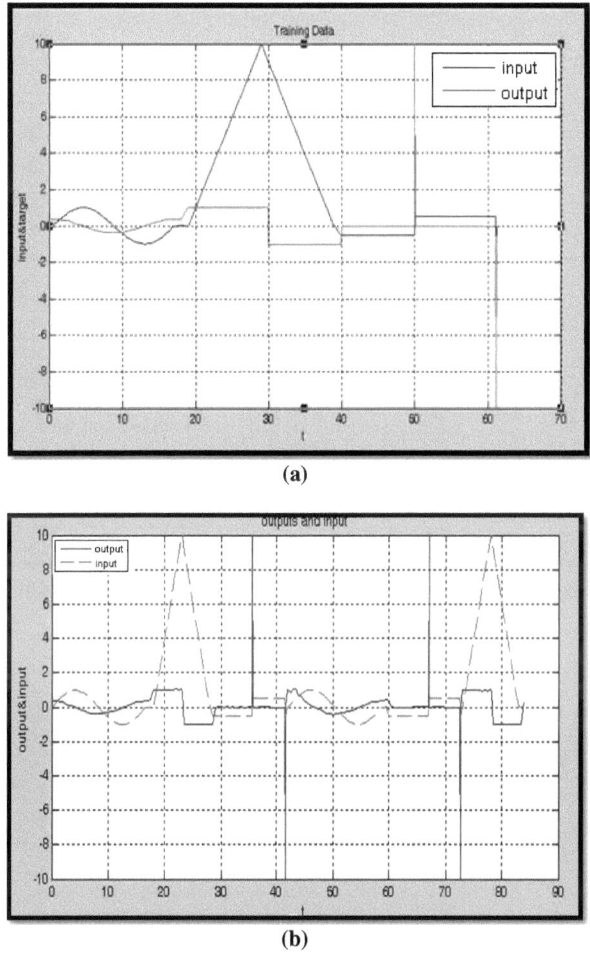

(a)

(b)

Figure (3-11): Many input waveforms (a) training signals (RC input/output) and (b) test signals (network input/output)

The final design of the neural network has 12 neurons in the first recurrent layer, 22 neurons in the second recurrent layer and one neuron in the output layer. By MATLAB, the training function is 'traingdx' and the performance function is 'mse', goal=0.00001 and no. of epochs=70000. The performance goal was eventually met at epoch

69977 (this big number because Elman NN is slower as compared to the feed-forward NN).

The final structure of Elman NN that obtained after is shown in Fig (3-12). And the recurrent connections of the hidden layers can be seen clearly.

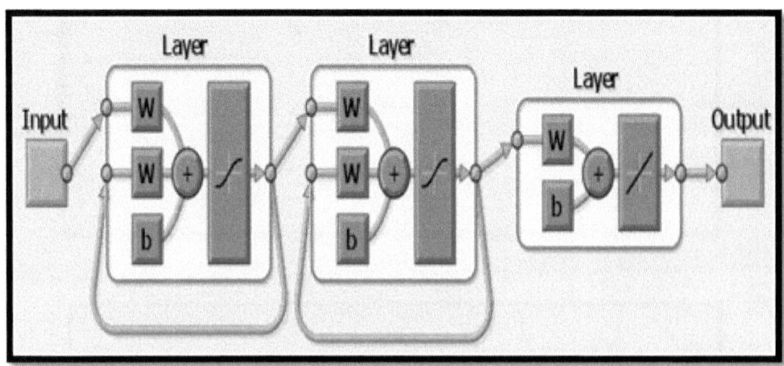

Figure (3-12): The final structure of Elman RNN after Training

Table (3-2) shows the comparison between some samples of the training and operating data. The Mean squared error is found to be (MSE $= \frac{\sum \text{Error}^2}{88} = 1.243*10^{-4}$), and RMS of error is (RMSE$=\sqrt{\frac{\sum \text{Error}^2}{N}}=0.01114899$) volt were N is the number of training set.

Table (3-2): Comparison between some samples of training data and neural network results

$$Vout = RC \; \frac{d(Vin-Vout)}{dt}$$

Seq.	Input sample voltage (volt) Vin	RC output voltage (volt) V_{RC}	NN output voltage (result) V_{NN}	$Error^2 = (V_{NN} - V_{RC})^2$
1	0	0.3729	0.3729	0
2	0.1837	0.3707	0.3707	0
3	0.3646	0.3513	0.3515	$4*10^{-8}$
4	0.5326	0.3194	0.3198	$1.6*10^{-7}$
5	0.6818	0.2763	0.2761	$4*10^{-8}$
6	0.8068	0.2233	0.2234	$1*10^{-8}$
7	0.9032	0.1624	0.1633	$8.1*10^{-7}$
8	0.9676	0.0958	0.0974	$2.56*10^{-6}$
9	0.9978	0.0258	0.0202	$3.136*10^{-5}$
10	0.9926	-0.0451	-0.0464	$1.69*10^{-6}$
11	0.9522	-0.1145	-0.1051	$8.836*10^{-5}$
12	0.8781	-0.1797	-0.1603	$3.7636*10^{-4}$
13	0.7729	-0.2387	-0.2436	$2.401*10^{-5}$
14	0.6403	-0.2891	-0.3037	$2.1316*10^{-4}$
15	0.4851	-0.3293	-0.3379	$7.396*10^{-5}$

3.5 <u>The Identification of the Dynamic System</u>:-

System identification is a modeling problem. Given a black box system, the goal of system identification is to develop a mathematical model to describe the relation between the input and output of the unknown system. If the system under consideration is memory less, the implication is that the output of this system is a function of presenting input only and bears no relation to past input. In this situation, the system identification problem becomes a function approximation problem. Figure (3-13) shows the structure of dynamic system identification.

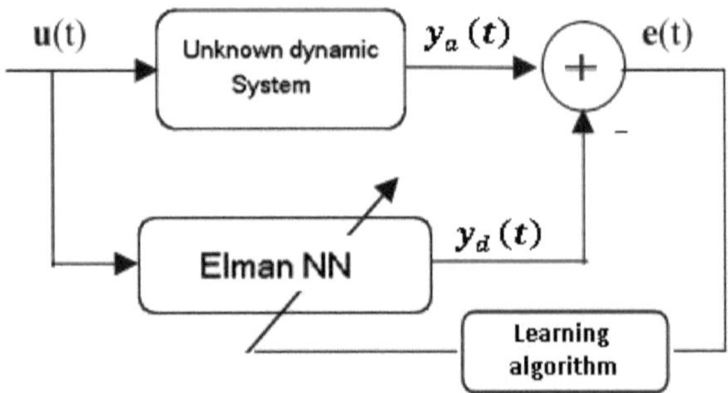

Figure (3-13): The structure of dynamic system identification [20].

The identification model is designed to minimize the error between expected and actual outputs [7, 20]

It is obvious now that any circuit can be trained by this designed RNN. And then the RNN works the same as the circuit. There are many advantages when having a redundant circuit. One of them is to find the fault when happens in the real circuit. Another advantage is to make compensation when there is a drift in the output.

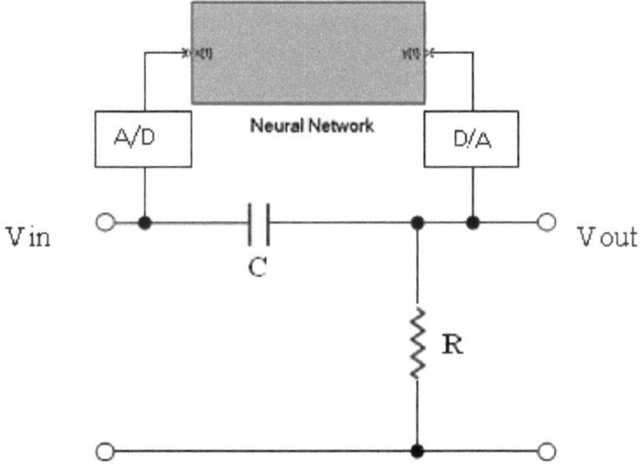

Figure (3-14): Real time connection

With online learning, the mathematical dynamic model receives the same inputs as the real unknown system, and produces an output $y_d(t)$ to approximate the true output $y_a(t)$. The difference between these two quantities will then be feedback to update the mathematical model [7,20].

In big order of equation or in complex electronic circuits, the designer makes simple modifications to improve the performance of Elman network. These modifications will give new types of Elman neural network.

The significant advantages of this approach compared to the conventional methods such as:-

1- Real time compensation for component change with temperature.

2- Predictive control: this means that we can predict the performance of electronic circuit and also find the position of fault in electronic circuit.

3- Flexibility in control there is several ways to connect the controller with real circuit as shown in Fig (3-15) were figure (3-15a) represent the connection between NN and electronic circuit using D/A and A/D convertors, and figure (3-15b) represent the connection between NN and electronic circuit using general purpose FPGA lines.

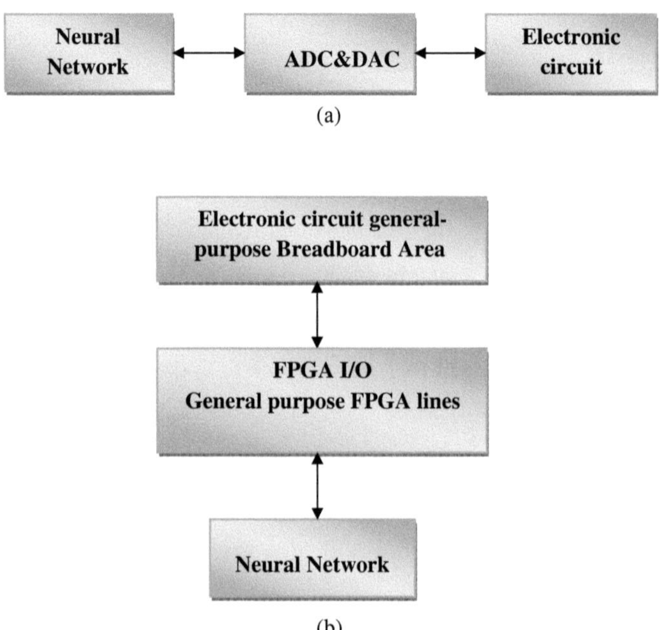

Figure (3-15): Ways of connecting NN and the electronic circuit

4- Need less hardware resources.

5- Any unknown dynamic system (electronic circuit) can be identified by Elman NN.

3.6 <u>Enhancement in RNN in order to Obtain more Accurate Results:-</u>

1- Fully connection of network had been used instead of partially connection.

A fully connected network is connected by all nodes among layers, unlike a partially connected network whose nodes only are connected between former and next layers. Partially connected network may be used as a filter. The partially connected network uses less weights memory; however, it was proved that less connections reduce the performance through simple experiments [37]. Figure (3-16) shows a fully and partially connections of NN.

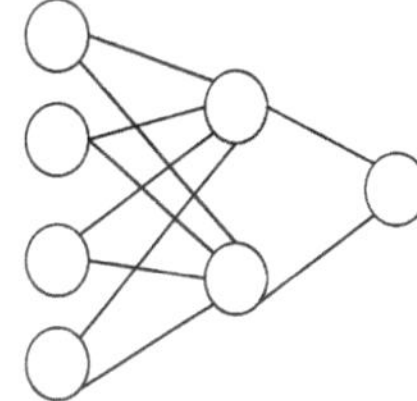 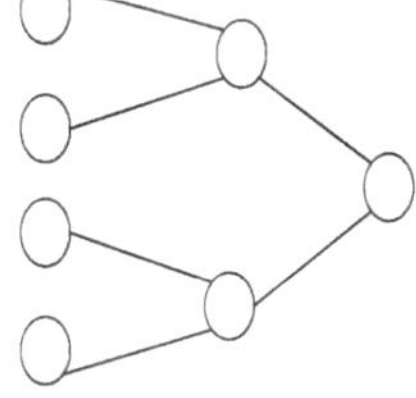

(a) Fully connected network   **(b) partially connected network**

Figure (3-16): Fully and partially connected network [37]

2- Using Elman NN whose performance can be improved by simple modification. Figure (3-17) shows the modified Elman NN

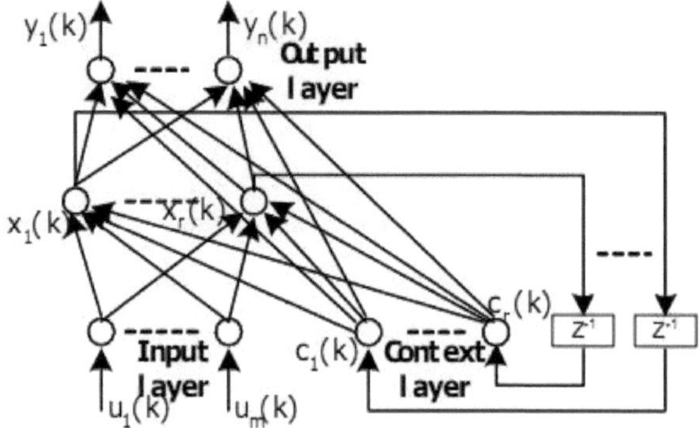

Figure (3-17): Modified Elman NN

Because the dynamic properties of Elman NN was improved to a certain degree. To improve the characteristics of the Elman NN a new partial RNN model was proposed as Fig (3-18) [20].

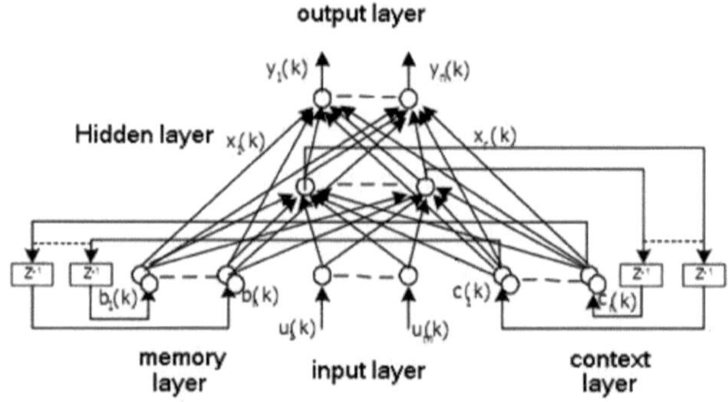

Figure (3-18): The structure of PID Elman neural network.

The new Elman network can utilize both the system static information and the dynamic information owing to its internal connection. It will further enforce the dynamic performance through the two context layers and dual loop feedback control [20].

3- Choose suitable number of samples that are used as training data because number of training data affects the learning of the network and then leads it go away from goal.

3.7 Neural Network for Logic Electronic Circuits:-

In this section, the designer used feed-forward neural network to design logic electronic circuit. The proposed designs are XOR and digital to analog converter.

3.7.1 XOR in neural network

The XOR truth table is shown in Table (3-3).

Table (3-3): The XOR Logical Operation

A	B	A XOR B
0	0	0
0	1	1
1	0	1
1	1	0

The XOR logical operation requires a slightly more complex neural network than the AND and OR operators. The neural networks presented so far have had only two layers an input layer and an output layer. More complex neural networks also include one or more hidden layers. The XOR operator requires a hidden layer [38]. XOR hidden layer consists of two neurons in the input layer and one output neuron. Figure (3-7(a,b)) show the structure of these layers.

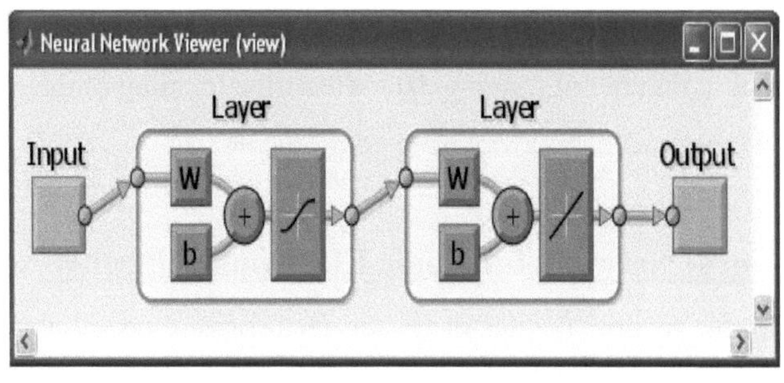

(a)

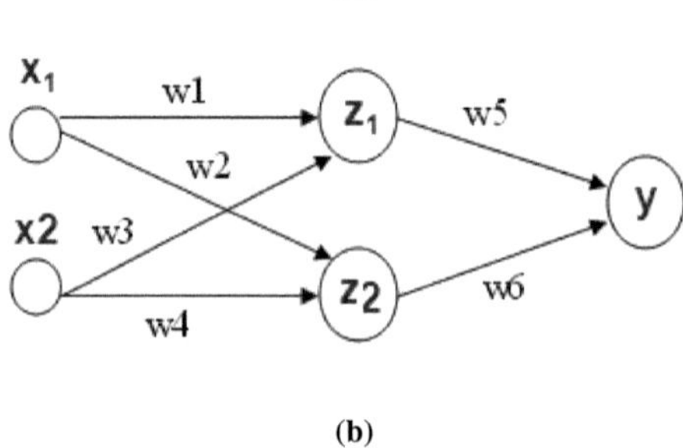

(b)

Figure (3-19): (a) The Structure of the network, (b) layers with neurons

Table(3-4) shows the weights obtained from training the designed network .

Table(3-4):- weights for XOR neural network

	weights
w1	1.9823
w2	-1.8263
w3	-1.1799
w4	1.4052
w5	1.1870
w6	1.1511

3.7.2 Digital to Analog Converter in Neural Network:-

Here, the design of 4-bit digital to analog convertor circuit is introduced. The circuit that represent 4-bit digital to analog convertor is shown in Figure (3-20).

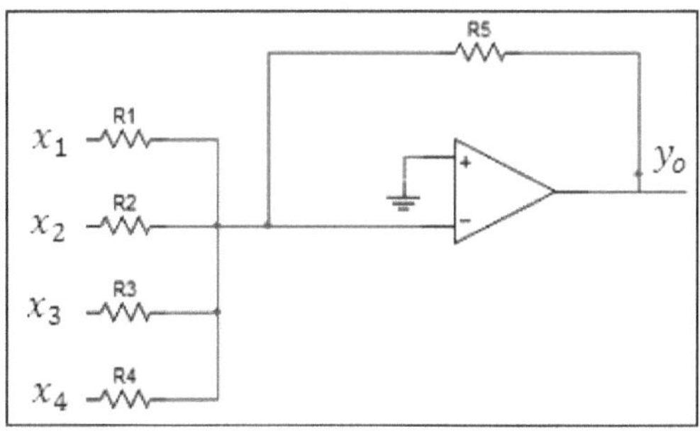

Figure (3-20): The Circuit of Digital to Analog Convertor

The DAC requires a hidden layer. DAC hidden layer consists of four neurons and one output neuron. Figure (3-21(a,b)) show the structure of these layers.

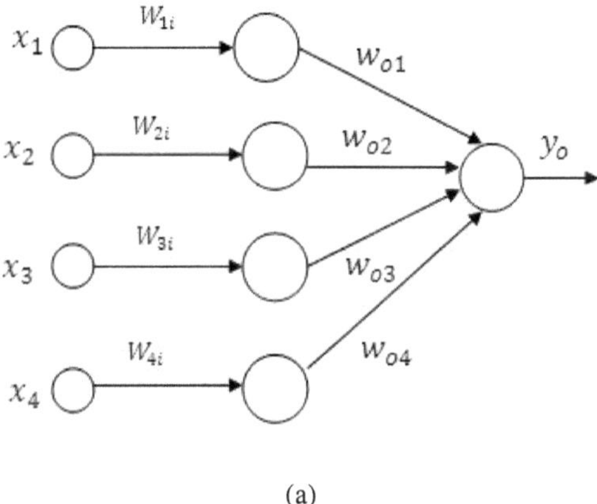

(a)

69

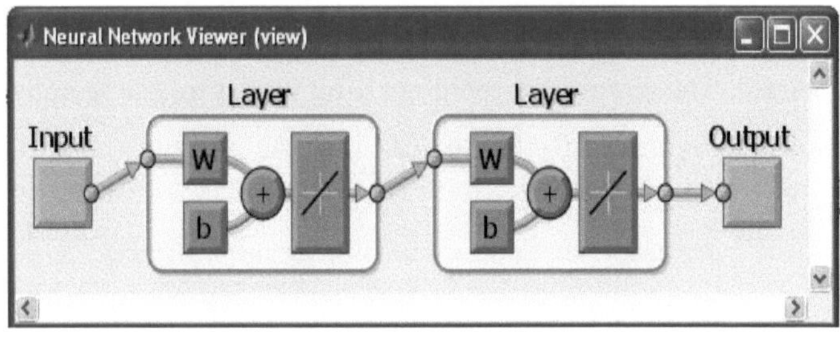

(b)

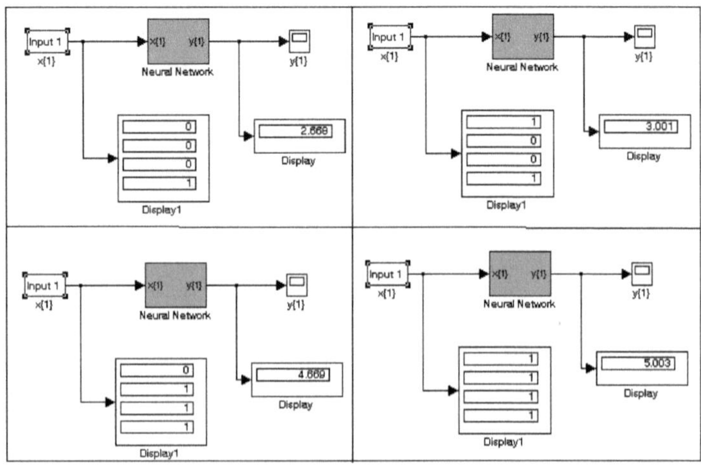

(c)

Figure (3-21): (a) layers with neurons, (b) The Structure of the network, (c) Test neural network with some inputs

Table (3-5) shows the comparison of the some samples of the training and operating data. The Mean square error squared is found to be (MSE = $\frac{\sum \text{Error}^2}{88}$=3.522*10^{-6}), and RMS of error is (RMSE=$\sqrt{\frac{\sum \text{Error}^2}{N}}$=1.87695*10^{-3}) volt.

Table (3-5): Comparison between Samples of training data and neural network results

Seq. (i)	Input sample x_1 x_2 x_3 x_4	output voltage $V_o = i * \dfrac{vcc}{2^n}$	NN output voltage V_{NN}	$Error^2 = (V_{NN} - V_o)^2$
1	0000	0	-0.0001929	$3.7*10^{-8}$
2	0001	0.3330	0.3333	$9*10^{-8}$
3	0010	0.6660	0.6678	$3.24*10^{-6}$
4	0011	1.0000	1.001	$1*10^{-6}$
5	0100	1.3330	1.333	0
6	0101	1.6660	1.667	$1*10^{-6}$
7	0110	2.0000	2.001	$1*10^{-6}$
8	0111	2.3330	2.335	$4*10^{-6}$
9	1000	2.6660	2.668	$4*10^{-6}$
10	1001	3.0000	3.001	$1*10^{-6}$
11	1010	3.3330	3.336	$9*10^{-6}$
12	1011	3.6660	3.669	$9*10^{-6}$
13	1100	4.0000	4.001	$1*10^{-6}$
14	1101	4.3330	4.335	$4*10^{-6}$
15	1110	4.6660	4.669	$9*10^{-6}$
16	1111	5.0000	5.003	$9*10^{-6}$

Where n represents the number of bits of the digital to analog convertor.

CHAPTER FOUR
FPGA Implementation of Neural Network Electronic Circuits

4.1 Introduction:-

This chapter presents an explanation about the FPGA Board Programmable with National Instrument LabVIEW and Xilinx ISE Tools which are integrated with NI ELVIS II/II and its components. It also contains an explanation of high throughput math functions and the difference between it and the LabVIEW numeric functions. After that , a new design of tan-sigmoid activation function using numeric and high throughput math function is presented. The designs of neural network for time series electronic circuits like RC circuit and the design of neural network for electronic logic circuits like XOR are introduced. In all designs, LabVIEW for FPGA VIs is used, and then downloaded onto the FPGA chips. Also, some designs were implemented using VHDL environments.

4.2 Educational Laboratory Virtual Instrumentation Suite (ELVIS):-

The NI Digital Electronics FPGA Board is a circuit development platform that was designed to help access to real-world signals and instrumentation for test, based on the XC3S500E Xilinx Spartan-3E field-programmable gate array (FPGA) architecture. Besides the FPGA, the board includes the necessary I/O that can be connected to the FPGA chip. Table (4-1) shows the necessary I/O available on the board. Those I/O devices are slide switches, LEDs, two digit seven-segment display, push-buttons, a rotary push-button knob and LEDs for one external clock, Digilent Pmod terminals for external attachments, USB download interface, and large breadboard area for digital electronics circuitry experimentation [39].

Table (4-1): I/O on the NI Digital Electronics FPGA Board [39]

I/O	Number
LEDs	8
DIP switches	8
Push buttons	4
Seven-segment LEDs	2
Pmod connectors	6
ADCs/DACs	2 AI ADCs and 4 AO DACs, 12-bit resolution
Encoder	1 rotary encoder with push-button shaft

The hardware components on the NI Digital Electronic FPGA Board are:

- FPGA Boot-Up Options.
- Breadboard Areas.
- Digilent Pmod Connectors.
- NIELVIS Connector   .

- Signal Descriptions.
- Slide Switches.
- Push Buttons.
- LEDs.
- Two Digit Seven-Segment Display.
- GPIO Lines.

- Rotary Push-Button Knob and LEDs.
- 50 MHz Onboard Oscillator.

Following Fig (4-1) shows NI Digital Electronics FPGA Board [39].

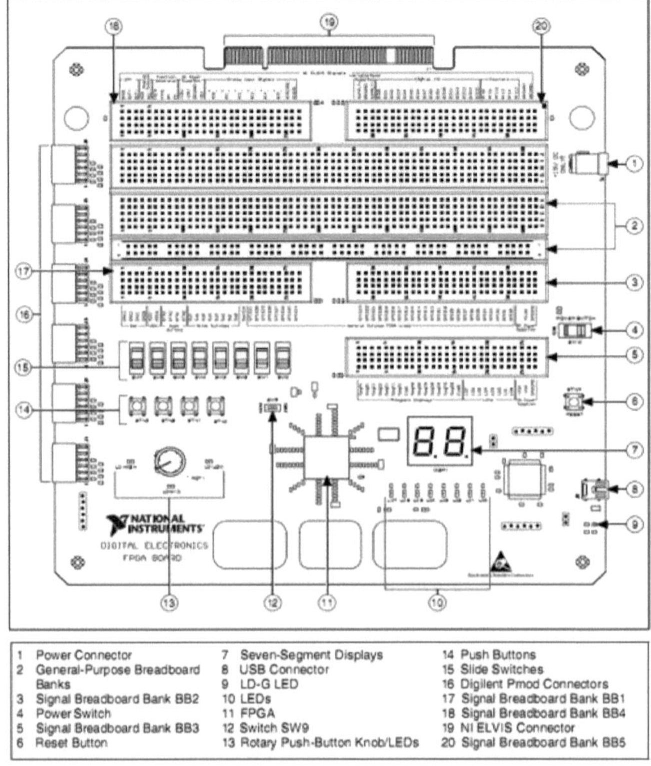

Figure (4-1): NI Digital Electronics FPGA Board [39].

NI ELVIS Connector: The NI Digital Electronics FPGA Board features a PCI type connector, shown in Fig (4-1), which plugs into an NI ELVIS II Series workstation when the NI Digital Electronics FPGA Board is used in NI ELVIS.

The NI Digital Electronics FPGA Board can work by two modes:-

- Stand-Alone Mode.
- NI ELVIS Mode.

In first mode, NI Digital Electronics FPGA Board connected only to a PC. Whereas in the second mode, the NI ELVIS Connector plugs into an NI ELVIS II Series workstation as shown in appendix B [39].

4.3 LabVIEW:-

National Instruments LabVIEW is a highly productive graphical programming environment that combines easy to use graphical developments with the flexibility of a powerful programming language. The graphical language is named "G" [29]. LabVIEW is a general purpose programming language used for developing projects graphically. It can also be called an Application-specific Development Environment (ADE). It is a highly interactive environment for rapid prototyping and incremental development of applications, from measurement and automation to real-time embedded and general purpose applications. National Instruments LabVIEW is a revolutionary programming language that depicts program code graphically rather than textually. LabVIEW excels in data acquisition and control, data analysis, and data presentation, while delivering the complete capabilities of a traditional programming language such as Microsoft Visual C [5]. One major benefit of using graphical programming rather than text-based languages is that one writes program codes simply by connecting icons. In addition, graphical programming solutions offer the performance and flexibility of text-based programming environments, but conceal many programming intricacies such as memory allocation and syntax. LabVIEW involves structured dataflow diagramming. It is, in fact, a much richer computational model than the control flow of popular text-based languages because it is inherently parallel, while C/C++ and VBA are not. They have to rely on library calls to operating system functions to achieve parallelism. Even a newcomer can design a highly parallel application. Also, LabVIEW is a dataflow programming language. Consequently, LabVIEW is a multitasking system capable of concurrently running multiple execution threads and multiple VI's (Virtual Instruments) [30].

LabVIEW programs/subroutines are called virtual instruments (VIs). Each VI contains the following three components:

- Front panel:- It serves as the user interface.

    LabVIEW utilizes a powerful Graphical User Interface (GUI). The Front panel constitutes one part of the program in which the GUI is developed and it is also acts as the user interface. Educators also can create interactive front panels in LabVIEW that interface to the FPGA, change the parameters, and see the results of their designs reflected immediately on the front panels [29, 30, 31].

- Blok diagram:- The block diagram is the background code. Thus, the block diagram constitutes the actual program part of LabVIEW. It is highly graphical, and no text codes are involved. The block diagram follows a similar idea as the 'data flow diagram' concept, in which the logic is presented as a diagram rather than as text code [29].

- Icon and connector pane:- Identifies the interface to the VI so that one can use the VI in another VI. A VI within another VI is called a sub VI. A sub VI corresponds to a subroutine in text-based programming languages [32].

4.4 LabVIEW FPGA

The NI LabVIEW FPGA Module can help to program a field-programmable gate array (FPGA) with a LabVIEW block diagram. Under the hood, the module uses code generation techniques to synthesize the graphical development environment to FPGA hardware. This block diagram approach to FPGA is well-suited for an intuitive depiction of the inherent parallelism that FPGAs provide.
Neural nets are parallel processors. They have data flowing in parallel lines simultaneously. LabVIEW has the unique ability to develop data flow diagrams that are highly parallel in structure. So LabVIEW seems to be a very effective approach for building neural nets [5].

Advantages of LabVIEW and LabVIEW FPGA Features
1- Graphic Design.
2- Fixed-point support.
3- Edit data-flow diagrams.
4- One major benefit of using graphical programming rather than text-based languages is that one writes program codes simply by connecting icons
5- Easy to use, no requirement on the knowledge of HDL [30].
6- The parallel processing of LabVIEW is well fit for FPGA [30].
7- VI can be easily tested before being embedded as a subroutine into a larger program [29].
8- LabVIEW is the relatively high speed of the compiler.
9- Visible on screen.
10- LabVIEW is a multitasking system capable of concurrently running multiple VI's.

11- Built-in VIs for Control, Filtering, and Signal Generation IP.

12- Implementation of FIFO, DMA and interrupt.

13- HDL Interface Node to import HDL code.

14- Component Level IP (CLIP) [30].

4.5 <u>Floating Point in LabVIEW FPGA:-</u>

Floating-point arithmetic is computationally expensive in hardware, and generally not supported on FPGA targets. Although LabVIEW FPGA does not support floating-point operations, we have found that it is possible to implement them using only the provided integer and Boolean types. Although the size required for a floating-point adder is an order of magnitude larger than that of a fixed-point adder

FPGA VIs support the following data types:

- 8-bit signed and unsigned integer numeric.
- 16-bit signed and unsigned integer numeric.
- 32-bit signed and unsigned integer numeric.
- 64-bit signed and unsigned integer numeric.
- Boolean.
- Fixed-point.

FPGA VIs also supports clusters and fixed-size, one-dimensional arrays of supported data types. Although these data types provide many benefits, under certain conditions their use can result in reduced performance or increased FPGA gate usage [33].

4.6 <u>High Throughput Math Functions (FPGA Module):-</u>

High Throughput Math functions can be used to perform high throughput math and analysis with fixed-point numbers on FPGA targets. These functions are similar to the LabVIEW Numeric functions but support higher throughput rates, handshaking terminals inside a single-cycle Timed Loop, input/output registers, and automatic pipelining. Elements of high throughput math function are shown in Fig (4-2).

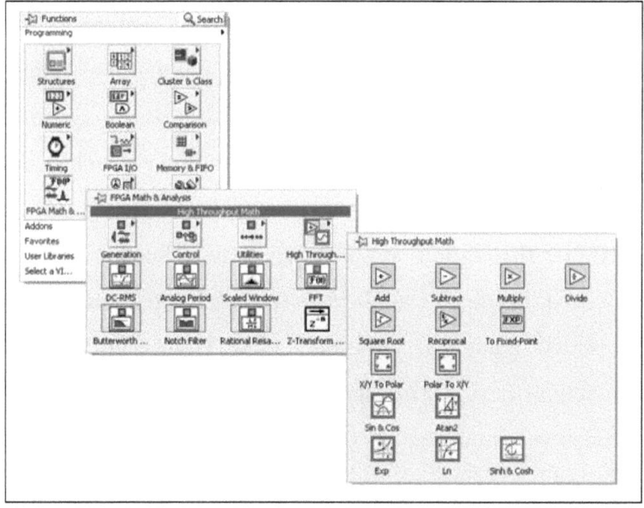

Figure (4-2): High Throughput Math Elements

 The High Throughput Math functions are different from the LabVIEW Numeric functions in the following ways:

- **Additional functions**: The High Throughput Math functions include trigonometric, logarithmic, and rectangular/polar coordinate conversion functions that support the fixed-point data type.

- **Universal single-cycle Timed Loop support**: Some Numeric functions cannot be placed inside a single-cycle Timed Loop. However, all High Throughput Math functions can be placed inside a single-cycle Timed Loop regardless of how many cycles that function takes to execute. Inside a single-cycle Timed Loop, these functions might display handshaking terminals. One use these terminals to ensure algorithms operate with only valid data. The High Throughput Math functions also feature ways to control the length of the combinatorial path to improve the clock rate the functions can achieve.

  - **Higher throughput support**: If pipeline Numeric functions are needed inside a single-cycle Timed Loop, these should be inserted manually. However, most High Throughput Math functions have a **Throughput** control. LabVIEW pipelines the function automatically to achieve the throughput rate you specify.

  - **Labeled terminals**: The High Throughput Math functions can be configured to display the encoding, word length, and integer word length of numeric terminals. You also can use the Context Help window to see configuration information about a function or wire.

National Instruments recommends using the LabVIEW Numeric functions unless there is a need to the benefits that the High

Throughput Math functions provide. The Numeric functions are easier to use when creating a VI and can execute on more platforms than the High Throughput Math functions can. For example, the High Throughput Math functions do not support NI real-time targets [30].

4.7 Neuron Activation Function Design:-

There are many ways to design the tan-sigmoid activation function of neurons, they differ from each other according to the components that used in the design. This section describes different ways to design transfer function of the activation function.

Firstly, tan-sigmoid function is illustrated in equation (4.1), can be designed by the traditional LabVIEW or by LabVIEW FPGA.

$$\tan(n) = \frac{2}{(1 + e^{-2n})} - 1 \dots \dots \dots \dots (4.1)$$

The design using traditional LabVIEW can be easily done as shown in Fig (4-3). But this design cannot be implemented on FPGA.

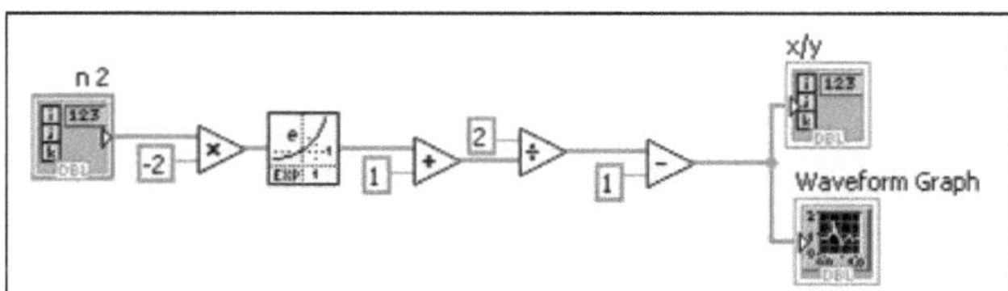

Figure (4-3): Tan-sigmoid LabVIEW program

Figure (4-4) shows the tan-sigmoid function that is obtained from the designed circuit shown in figure (4-3).

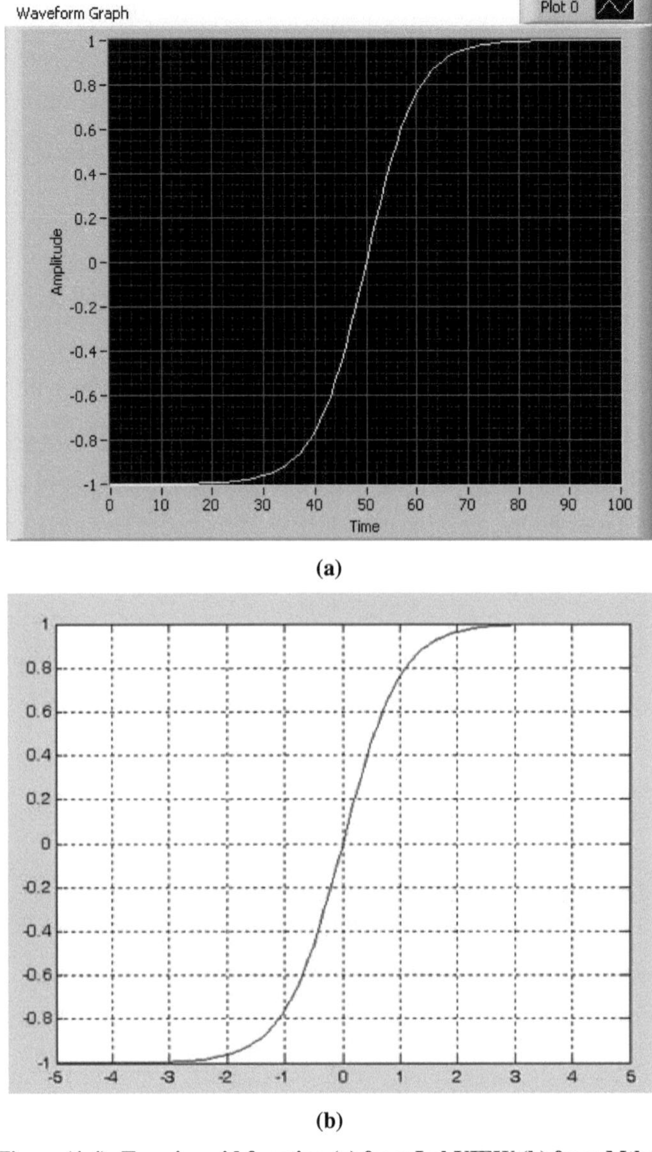

(a)

(b)

Figure (4-4): Tan-sigmoid function (a) from LabVIEW (b) from Mtlab

The design of figure (4-3) cannot be implemented in LabVIEW FPGA because:

1- LabVIEW FPGA work only with data type that listed in section (4.5)

2- The exponential function is not provided in the numeric function categories.

3- Also, the size of any array and connections must be fixed and other limitations must be considered when there is a need to design with LabVIEW FPGA.

On the other hand, inside LabVIEW FPGA tan-sigmoid function can be implemented by two ways:

- **Using Numeric Function:** by this way, the design of the approximation function using numerical computation can be implemented.

- **Using High Throughput Math Function:** Here the designed hardware for the function can be implemented with or without using the approximation of the function.

4.7.1 Design tan-Sigmoid Using the High Throughput Math Functions (FPGA Module)

As mentioned in chapter two, when someone need to use the High Throughput Math functions, the analysis and design must be in fixed-point numbers. And there will be no significant difference in performance specially for medium sized operations and projects.

In this section, high throughput math function will be used to design the tan-sigmoid transfer function. We made use of some benefits of the High Throughput Math functions provided by LabVIEW like the Exponential function in order to design transfer function without approximations as shown in the Fig (4-5):

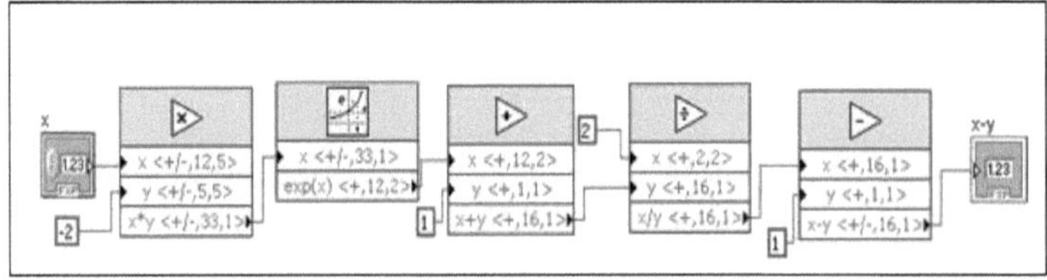

Figure (4-5): Design Tan-sigmoid without approximation using high throughput math function

After compiling the program of figure (4-5), the consumed resources that obtained to implement tan-sigmoid in high throughput (HT) are listed inn table (4-2).

Table (4-2): The Consumed Resources to implement Tan-sigmoid in HT

```
Device utilization summary:
---------------------------
Selected Device : 3s500eft256-5
 Number of Slices:                     766   out of    4656     16%
 Number of Slice Flip Flops:           798   out of    9312      8%
 Number of 4 input LUTs:              1299   out of    9312     13%
    Number used as logic:            1290
    Number used as Shift registers:     9
 Number of IOs:                        126
 Number of bonded IOBs:                112   out of     190     58%
 Number of MULT18X18SIOs:                1   out of      20      5%
 Number of GCLKs:                        2   out of      24      8%
Timing Summary:
---------------
```

```
Speed Grade: -5
   Minimum period: 10.107ns (Maximum Frequency: 98.941MHz)
   Minimum input arrival time before clock: 3.316ns
   Maximum output required time after clock: 0.871ns
```

Unfortunately, the length of the integer word of the HT exponential function cannot be changed as it is obvious from Fig (4-6). While for other functions, data type (i.e. integer word length) can be modified using the property menu.

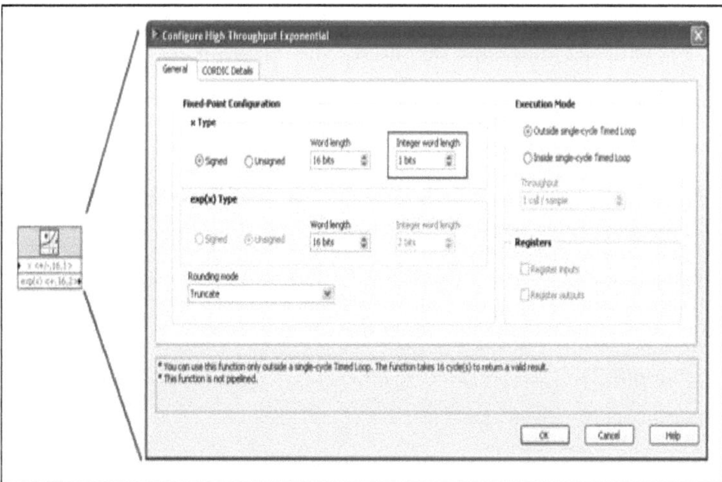

Figure (4-6): Exponential function configuration

Another approach was used to design the tan-sigmoid activation function using the available high throughput function. Second order approximation was used for that.

Equation (2.3) from chapter two will be rewritten here, which is a second order nonlinear function of approximated tan-sigmoid [18].

$$f(n) = \begin{cases} n\,(B - g.n) & for\ 0 \leq n \leq L \\ n\,(B + g.n) & for\ -L \leq n < 0 \end{cases} \quad ...(4.2)$$

Where B and g: slop and gain respectively of the nonlinear function.

$$B = 2/L \ ...... (4.3)$$

$$g = 1/L^2 \ ...... (4.4)$$

To implement the previous equation in LabVIEW FPGA, the case condition element can be used by setting (L=1). Case structure, has one or more sub diagrams or cases. In this design as shown in figure (4-6), two sub diagrams are used. Sub diagram in figure (4-7a) is executed when the switch is true and sub diagram Figure (4-7b) is executed when the switch is false. The value wired to the selector terminal determines which case is executed. The switch can be Boolean, string, integer, or enumerated type. In true case, the implementation of the first part of equation (4.2) can be seen. Then, Fig (4-7b) represents the program for the second part of equation equation(4.2)

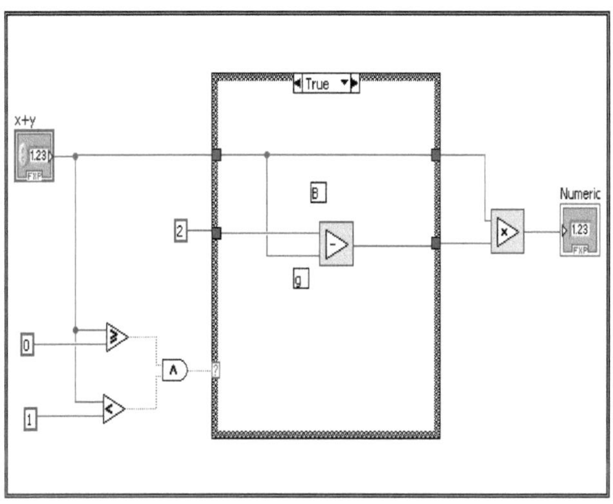

(a)

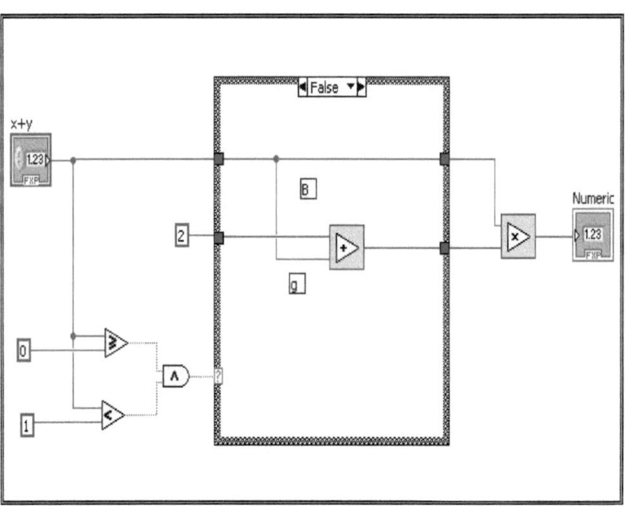

Figure (4-6): Design of Tan-sigmoid with Approximation using HT (a) True condition, (b) False condition.

Table (4-3) illustrates the consumed resources that are obtained to implement the approximated tan-sigmoid in HT when using function (4.2):

Table (4-3): Consumed Resources to Implement the Approximated tan-sigmoid in HT

```
Device utilization summary:
---------------------------
Selected Device : 3s500eft256-5
 Number of Slices:              435   out of   4656     9%
 Number of Slice Flip Flops:    624   out of   9312     6%
 Number of 4 input LUTs:        533   out of   9312     5%

Timing Summary:
---------------
Speed Grade: -5
   Minimum period: 7.517ns (Maximum Frequency: 133.039MHz)
   Minimum input arrival time before clock: 3.316ns
   Maximum output required time after clock: 0.871ns
```

4.7.2 Design Tan-sigmoid Using Numeric Functions:-

Here, instead of using high throughput functions to program equation (4.2), numeric function were used to approximate the tan-sigmoid activation function. Figure (4-8) shows the program for that.

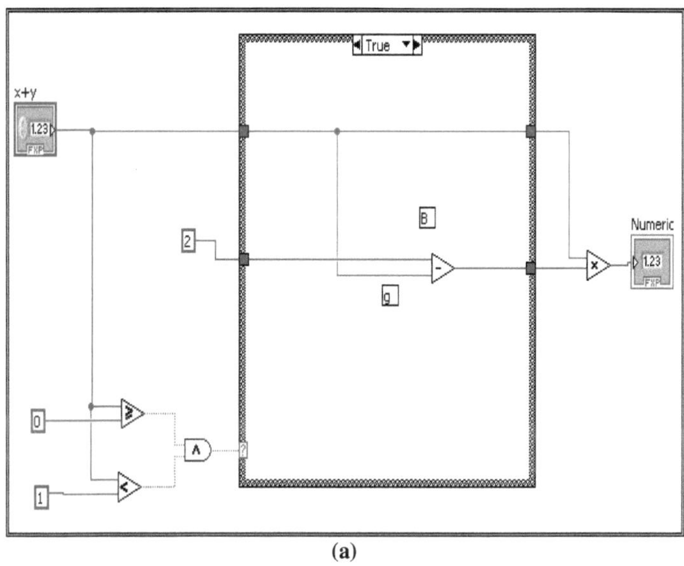

(a)

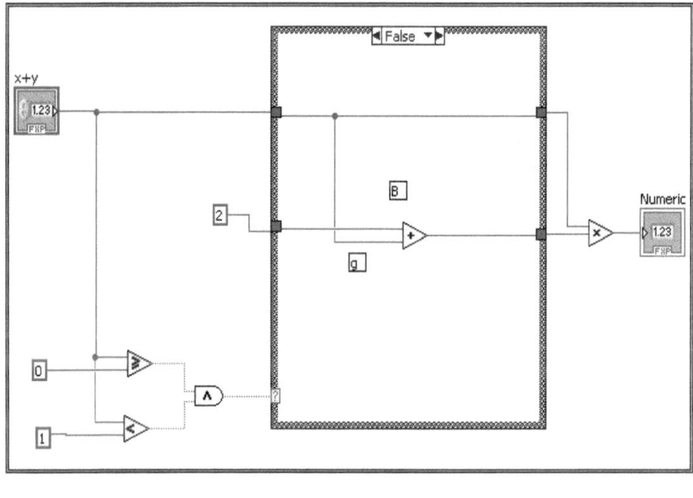

(b)

Figure(4-8): Design of Tan-sigmoid with Approximation using Numeric Function (a) True condition,(b) False condition

Table (4-4) illustrates the consumed resources that are obtained to implement the approximated tan-sigmoid when using numeric functions.

Table (4-4): Consumed Resources to Implement the Approximated Tan-sigmoid using numeric function

```
Device utilization summary:
---------------------------
Selected Device : 3s500eft256-5
 Number of Slices:                 436   out of    4656      9%
 Number of Slice Flip Flops:       623   out of    9312      6%
 Number of 4 input LUTs:           529   out of    9312      5%
    Number used as logic:          524

Timing Summary:
---------------
Speed Grade: -5
   Minimum period: 7.517ns (Maximum Frequency: 133.039MHz)
   Minimum input arrival time before clock: 3.316ns
   Maximum output required time after clock: 0.871ns
```

In this thesis, numeric functions rather than high throughput math functions are used for many reasons:-

- National Instrument recommends to use the LabVIEW Numeric functions unless need the benefits that the High Throughput Math functions provide [40].

- The Numeric functions are easier to use when creating a VI and can execute on more platforms than the High Throughput Math functions can. For example, the High Throughput Math functions do not support NI real-time targets [40], and it uses one data type (fixed point only), on the other hand, numeric functions work with all data types that are supported by LabVIEW FPGA.

The comparison between using high throughput math function and using numeric function is shown in Table (4-5)

Table (4-5): Comparison between the results that obtained to design Tan-sigmoid

Device utilization and time summary	tan-sig using HT	Approx. tan-sig using HT	Approx. tan-sig numeric function
No. of Slices	16%(766 out of 4656)	9%(435 out of 4656)	9%(463 out of 4656)
No. of slices FF	8% (798 out of 9312)	6% (624 out of 9312)	6% (623 out of 9312)
No. of 4 input LUTs	13% (1299 out of 9312)	5% (533 out of 9312)	5% (529 out of 9312)
Minimum period	10.107ns	7.517ns	7.517ns
Maximum frequency	98.941MHz	133.039MHz	133.039MHz

From the table, there is no big significant difference when using numeric functions or HT functions. HT version use less resources, but only because it is configured with smallest data type (fixed point only).

Finally, the designed tan-sigmoid using numeric functions is used to design neuron as shown in figure(4-9).

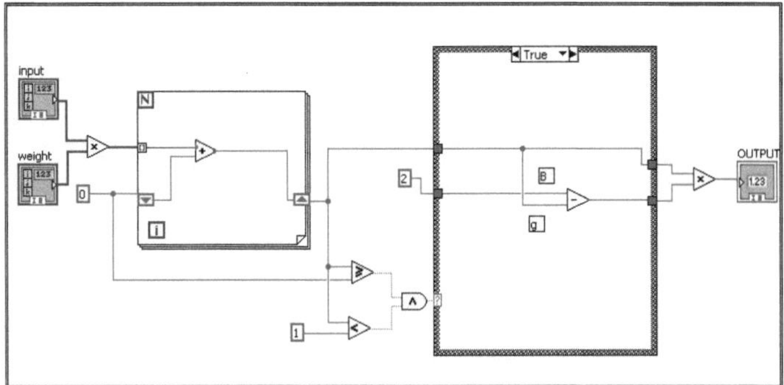

Figure (4-9): Neuron with tan-sigmoid function

4.8 <u>Neural Network Implementation on FPGA for Electronic Logical Circuits</u>:-

In this section, the design of a neural networks for logical and hybrid circuits like XOR and D/A convertor and their implementation on FPGA using LabVIEW FPGA environments are introduced.

4.8.1 LabVIEW FPGA Design for XOR Neural Network with and without:-

To design the XOR circuit with neural network, a neuron has to be designed as shown in Fig (4-10).

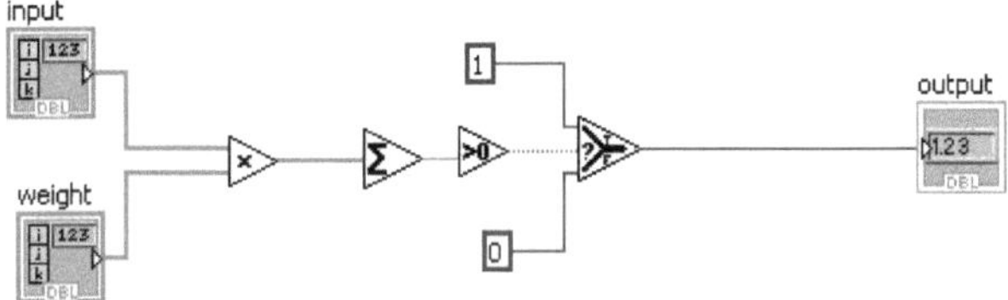

Figure (4-10): Neuron in LabVIEW

After building a VI, the designer needs to build the icon and the connector pane so that this VI can be used as a sub-VI. The connector that is used as neuron is shown in Fig (4-11).

An icon is a graphical representation of a VI. It can contain text, images, or a combination of both. If one uses a VI as a sub-VI, the icon identifies the sub-VI on the block diagram of the VI [5].

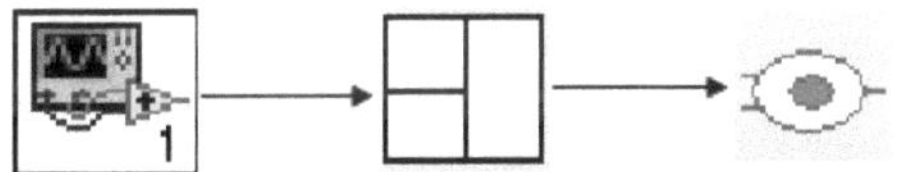

Figure (4-11): Explanation of Icons and Connector panes

Neural network XOR circuit shown in Fig (4-11) consists of two input neurons and one output neuron.

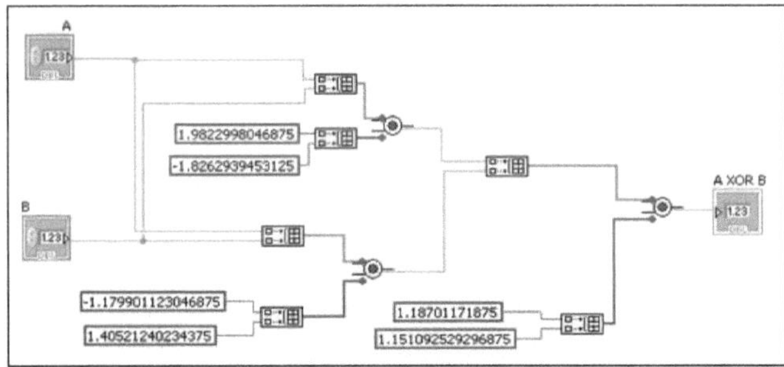

Figure (4.12): Network of XOR Circuit in LabVIEW

In LabVIEW environment, all elements are provided. On the other hand, in LabVIEW FPGA environment there are some limitations such as all matrices must be of fixed sizes, FPGA VIs only support (Integer, Boolean, Fixed-point) data format while floating point format is not supported, also some functions are not supported on FPGA targets like Add array elements which is shown in Fig (4-12) .

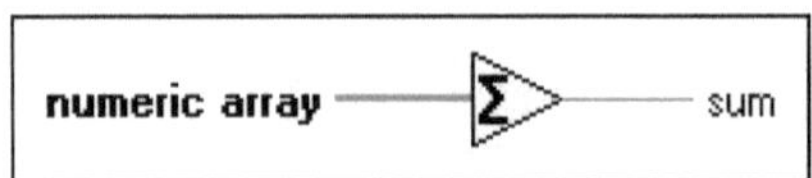

Figure (4-13): Add Array Element

The neuron that had been designed for XOR circuit by LabVIEW FPGA is shown in Fig (4-14). It is obvious that Add Array element function is not used in LabVIEW FPGA because it is only available in LabVIEW environments. Therefore, instead we used a summation element with shift register . Also, data is presented by fixed size integers.

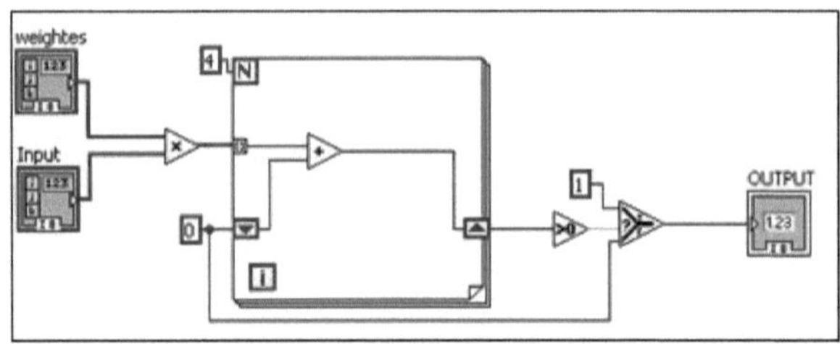

Figure (4-14): Neuron for XOR circuit in LabVIEW FPGA

After building the sub _VI which represents a neuron, this sub-VI can be used to build the network that woks as XOR circuit which is shown in Fig (4-15)

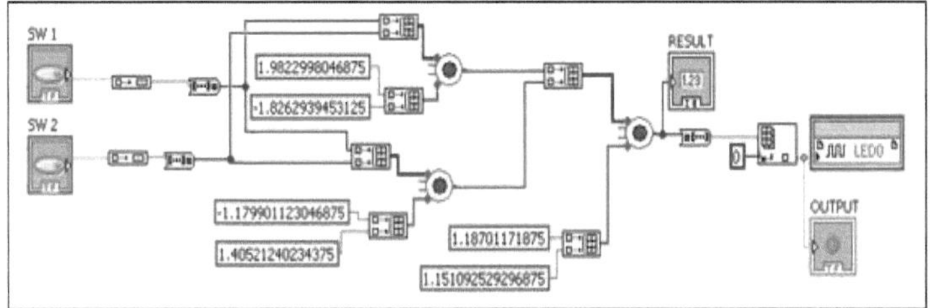

Figure (4-15): Neural Network of XOR circuit in LabVIEW FPGA with Boolean Switch

Front panel objects appear as terminals on the block diagram. The terminals represent the data type of the control (input) or indicator (output). Therefore, the circuit was tested by the host computer and the results are shown in Fig (4-15). The front panel controls or indicators can be configured to appear as icons or data type terminals on the block diagram. By default, front panel objects appear as icon terminals.

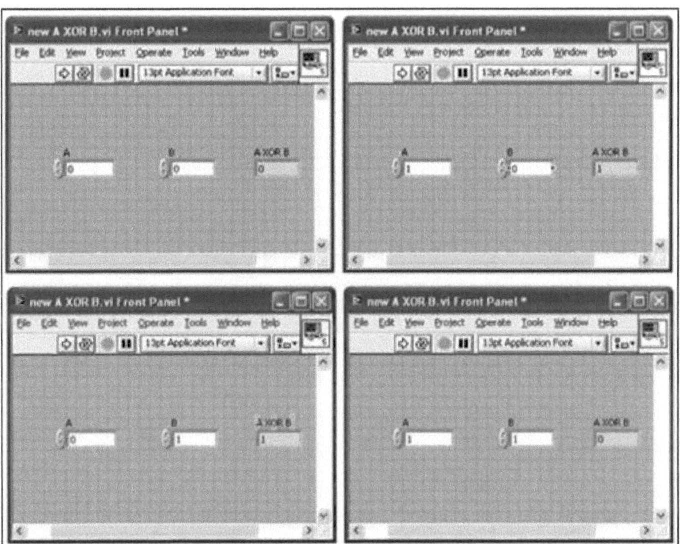

Figure (4-16): Test of the XOR circuit by host computer

The editor has a rich set of operations to quickly create elaborate user interfaces by direct manipulation. The fact that every module, or VI, has a user interface means that interactive testing is simple to do at each step, without writing any extra code. The fraction of an application that has to be completed before meaningful testing can take place is much smaller in LabVIEW than in traditional programming tools, making design much faster [6]. Figure (4-17) shows the block diagram of the neural network for XOR for which is represented on the FPGA device and the results of this circuit on the front panel of the ELVIS training system.

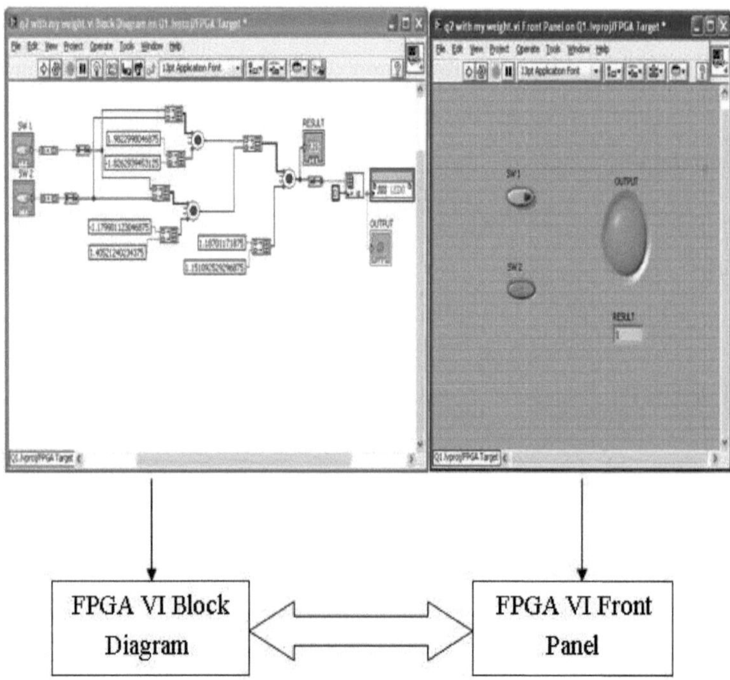

Figure (4-17): FPGA VI of XOR Neural Network

Figure (4-18): XOR neural network in ELVIS

After compilation of the VI, it is noted that the consumed resources are as tabulated in table (4-6)

Table (4-6): Consumed Resources to Implement the XOR in Elvis system

```
Device utilization summary:
-----------------------------
Selected Device : 3s500eft256-5
 Number of Slices:                        716   out of   4656    15%
 Number of Slice Flip Flops:              988   out of   9312    10%
 Number of 4 input LUTs:                 1047   out of   9312    11%
     Number used as logic:               1042
     Number used as Shift registers:        5
 Number of IOs:                           126
 Number of bonded IOBs:                   112   out of    190    58%
     IOB Flip Flops:                        1
 Number of MULT18X18SIOs:                   3   out of     20    15%
 Number of GCLKs:                           2   out of     24     8%
Timing Summary:
---------------
Speed Grade: -5
    Minimum period: 6.337ns (Maximum Frequency: 157.802MHz)
    Minimum input arrival time before clock: 3.316ns
    Maximum output required time after clock: 4.040ns
```

I/O switches and LEDs when added to the design, the network design becomes as shown in figure (4-19) with the results shown in figure (4-20), and the consumed resources become as listed in table (4-8).

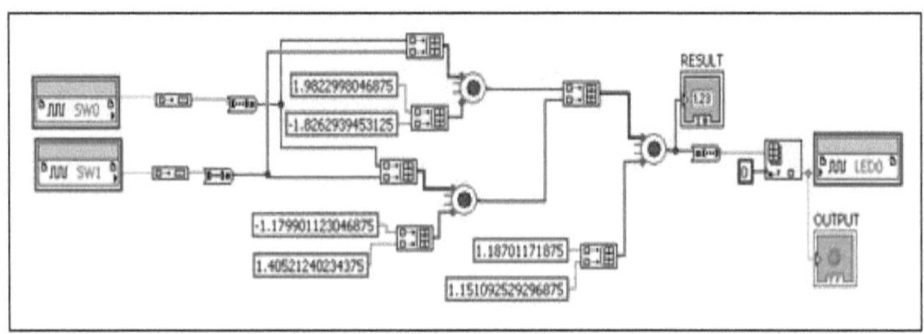

Figure (4-19): Neural Network of XOR circuit with I/O FPGA switch

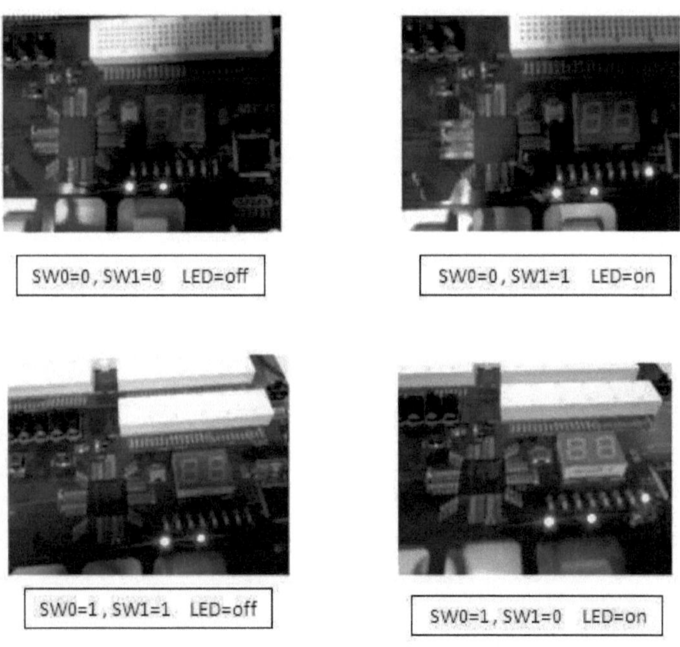

Figure (4-20): XOR neural network in ELVIS with switches

Table (4-7) illustrates the consumed resources that obtained to implement the XOR circuit with I/O of FPGA.

Table (4-7): Consumed Resources to Implement the XOR with switch

```
Device utilization summary:
---------------------------
Selected Device : 3s500eft256-5
 Number of Slices:             712   out of   4656     15%
 Number of Slice Flip Flops:   986   out of   9312     10%
 Number of 4 input LUTs:      1038   out of   9312     11%
 Number of IOs:                126
 Number of bonded IOBs:        114   out of    190     60%
    IOB Flip Flops:              1
 Number of MULT18X18SIOs:        3   out of     20     15%
 Number of GCLKs:                2   out of     24      8%

Timing Summary:
---------------
Speed Grade: -5
   Minimum period: 6.363ns (Maximum Frequency: 157.165MHz)
   Minimum input arrival time before clock: 3.316ns
   Maximum output required time after clock: 4.040ns
```

4.9 RC circuit design using proposed Elman Neural Network in LabVIEW FPGA Environments:-

LabVIEW has many interesting features that make it a very useful tool for building neural nets. LabVIEW programs are called Virtual Instruments (VI), because they appear very similar to actual laboratory instruments [6].

In this section, circuit design of time series neural network using LabVIEW FPGA will be seen. First of all, the activation functions of neurons using high throughput math functions with approximation and without approximation and also using numeric functions are all considered and the optimal one is chosen in the design Elman neural network.

4.9.1 Configuration of the Neural Network

Elman neural network that consists of three layers (two recurrent layers and one output layer), will be considered in the design. Through the design, many difficulties appeared such as, LabVIEW FPGA supports one dimensional fixed size array. If the dimension changes during the work, dealing with more than one size of arrays generates warning error from the LabVIEW software. Figure (3-6) in chapter three shows the structure of Elman neural network which is to be implemented in LabVIEW FPGA environment.

Firstly, all the arrays that will be used in the design have to be specified of fixed sizes. Figure (4-20) shows how to set the size of an array and to choose the type of the size. On the menu, one can go through (block diagram programming>>Numeric>> conversion) and then choose the desired type, or right click on the (Indicator>>representation) and then choose the required representation. Also, from front panel right click on (Indicator>> set dimension size) and then choose the size and set the required number of the array elements.

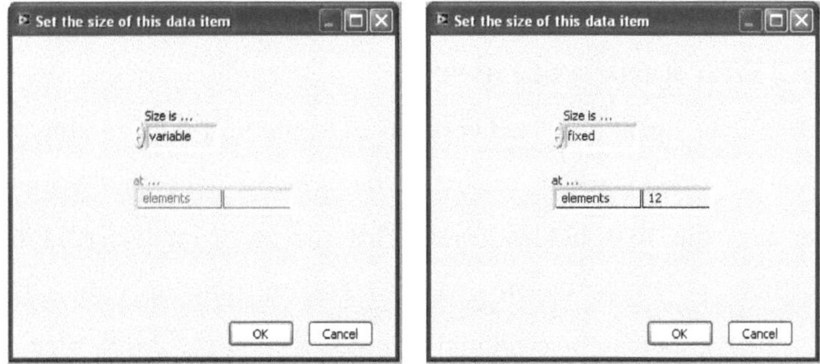

Figure (4-21): Set the Size and the Number of Elements

Form Fig (4-21), there are (12 neurons) in the first recurrent layer, (22neurons) in the second, and finally, one output neuron.

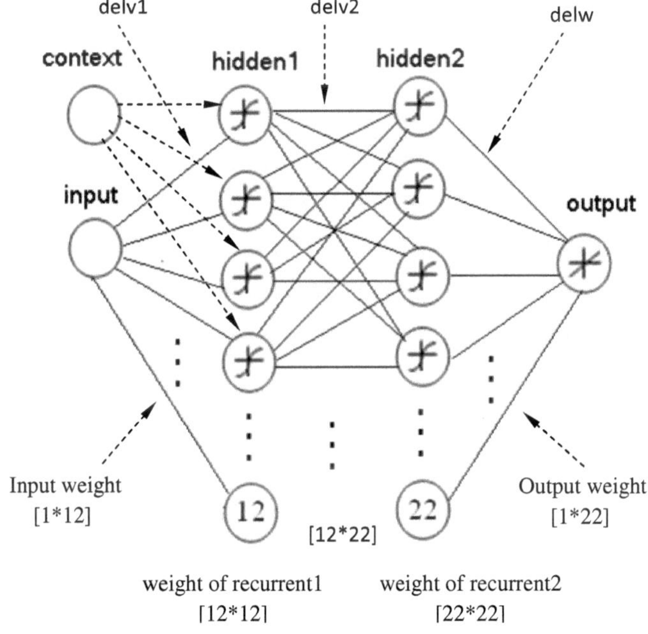

Figure (4-22): The Designed Neural Network

4.9.2 First Recurrent Layer

The input signal to the network has to be multiplied by an array of dimension [1*12], which represents the weights between the input layer and the first hidden layer. This part was implemented by LabVIEW FPGA as shown in Fig (4-23a). In order to reduce the consumed resources, one multiplier inside a loop can fulfil what is required, and that is shown in figure (4-23b). The consumed resources for this configuration are shown in table (4-8)

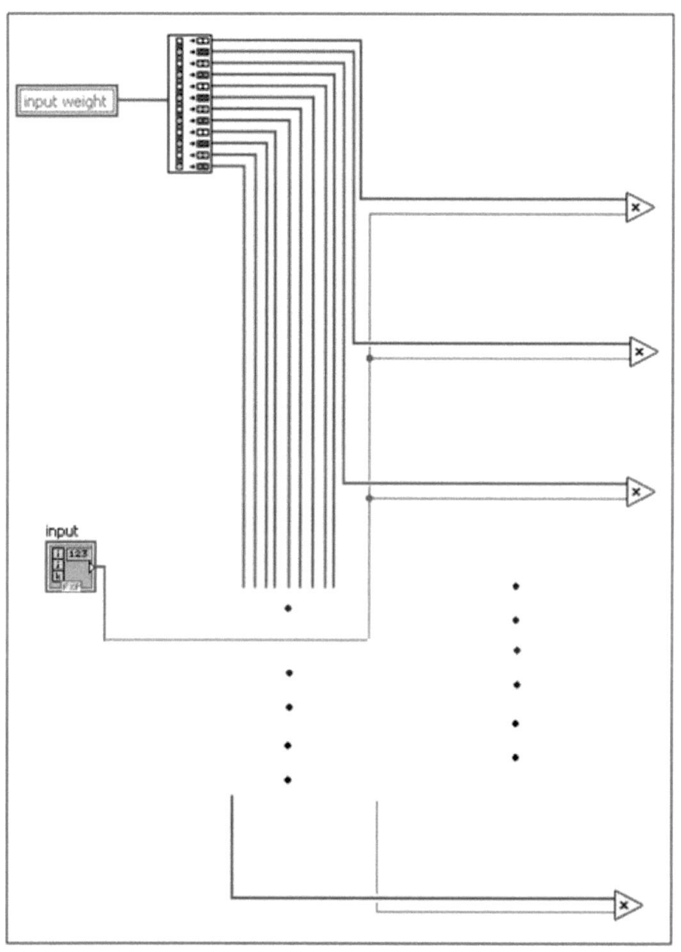

(a)

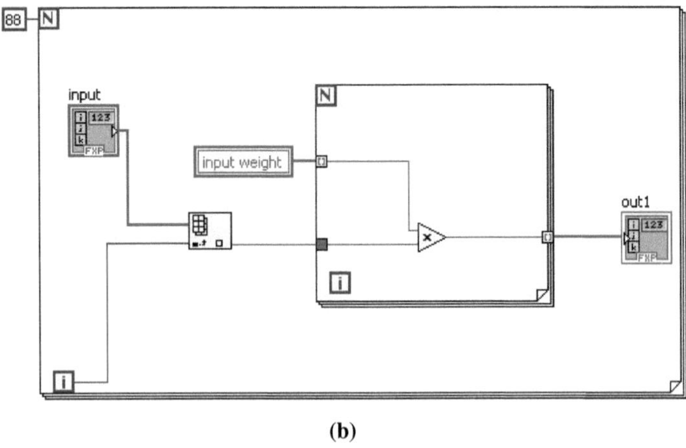

(b)

Figure (4-23):- Two VIs for the connection of input and first hidden layer (a) VI for Input and First Recurrent layer in LabVIEW FPGA, (b) looping the input to the First Recurrent layer in LabVIEW FPGA VI for less consumed resources

Table (4-8): Consumed Resources of the input layer and its connections

```
Device utilization summary:
----------------------------
Selected Device : 3s500eft256-5
 Number of Slices:                   1618   out of    4656    34%
 Number of Slice Flip Flops:         2058   out of    9312    22%
 Number of 4 input LUTs:             2005   out of    9312    21%
    Number used as logic:            2000
    Number used as Shift registers:     5
 Number of IOs:                       126
 Number of bonded IOBs:               112   out of     190    58%
 Number of MULT18X18SIOs:               1   out of      20     5%
 Number of GCLKs:                       2   out of      24     8%
Timing Summary:
---------------
Speed Grade: -5
   Minimum period: 7.333ns (Maximum Frequency: 136.375MHz)
   Minimum input arrival time before clock: 3.316ns
   Maximum output required time after clock: 0.871ns
```

The results coming out from the circuit of Fig (4-22) are used as inputs to the first recurrent layer. Other inputs to this recurrent layer come from the context layer whose outputs have to be multiplied by $[12 * 12]$ array which represents the weights of feed-back connection. The parallel and sequential designs for the first recurrent layer are shown in figure (4-24) a and b respectively.

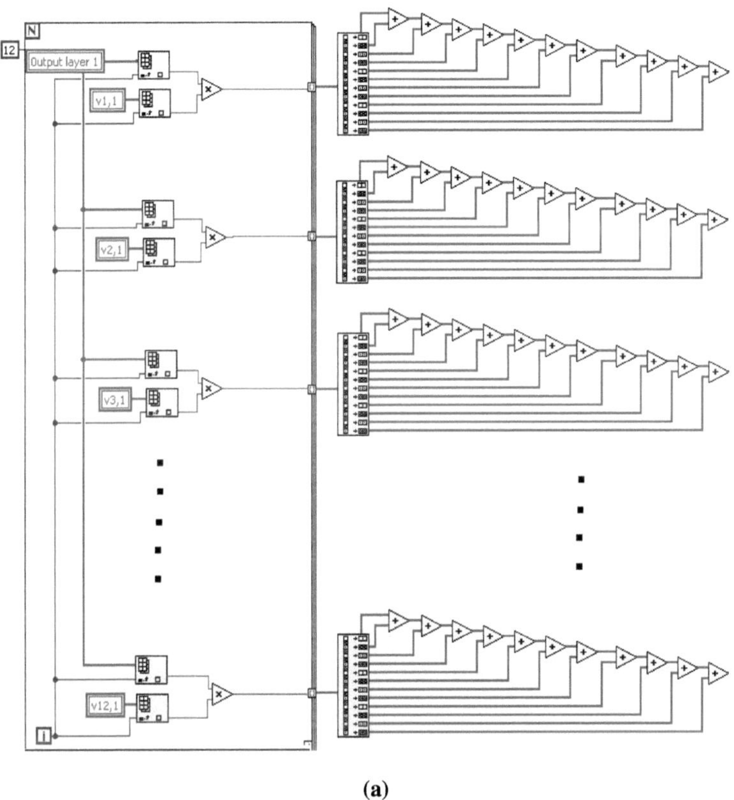

(a)

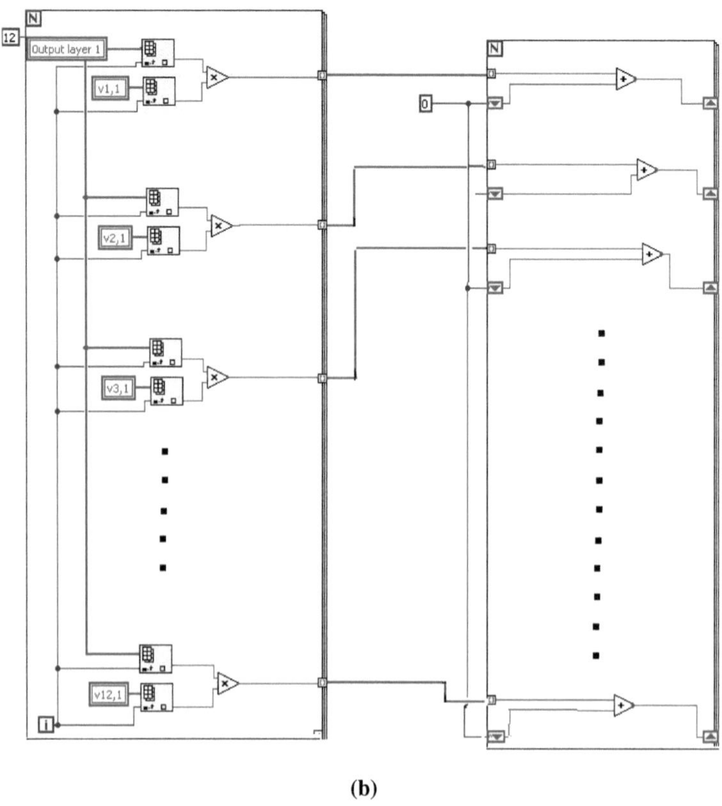

(b)

Figure (4-24): (a) Design of the First Recurrent layer in LabVIEW
FPGA VI (b) Looping the design of first Recurrent layer in LabVIEW
FPGA VI

Then, this recurrent layer adds as another adds as another input the bias array of size $[1 * 12]$. The resultant sum is then fed to the activation function that was previously designed and shown in Fig (4-24). Figure (4-25) shows the complete design of the first recurrent layer. Activation function VI is considered as a sub VI called by this layer. Summery of the consumed resources of the FPGA chip for the first recurrent layer is shown in table (4-9). Many designs was tried, in each one, a modification or more was tried in order to reduce the consumed resources. At the end of this chapter, there will be the ways that were followed to make enhancements to the design.

Table (4-9): Consumed Resources of first recurrent layer

Device utilization	Design number					
Sea. of reduce	1	2	3	4	5	6
Number of Slices (4656)	800%	694%	595%	444%	239%	193%
No. of slices FF (9312)	302%	370%	399%	263%	182%	148%
No. of 4 input LUTs(4656)	574%	451%	281%	246%	129%	119%
No. of multiplier	340%	105%	100%	100%	75%	70%
Min. period (ns)	20.676	26.33	13.74	13.74	8.996	8.996
Max. frequency (MHz)	48.36	37.97	72.76	72.76	111.157	111.157

4.9.3 Second Recurrent Layer

This layer is similar to the first recurrent layer except that it has (22 neurons) instead of 12 neurons. This layer multiplies the outputs of the first recurrent layer by $[12 * 22]$ array weights. The second type of inputs for this layer is from context layer whose outputs are multiplied by $[22 * 22]$ array of weights. Figure (4-26) shows the VI for the second recurrent layer. The consumed resources can be found in table(4-10).

Table(4-10):- Consumed Resources of second recurrent layer

```
Device utilization summary:
-----------------------------
Selected Device : 3s500eft256-5
 Number of Slices:                29940 out of    4656   643%
 Number of Slice Flip Flops:      43479  out of   9312   466%
 Number of 4 input LUTs:          26946  out of   9312   289%
    Number used as logic:         26941
    Number used as Shift registers:   5
 Number of IOs:                   126
 Number of bonded IOBs:           112   out of    190    58%
 Number of MULT18X18SIOs:          20   out of     20   100%
 Number of GCLKs:                   2   out of     24     8%
Timing Summary:
---------------
Speed Grade: -5
   Minimum period: 13.369ns (Maximum Frequency: 74.79MHz)
   Minimum input arrival time before clock: 3.316ns
   Maximum output required time after clock: 0.871ns
```

4.9.4 Output Layer:-

This layer multiplies the output of the second recurrent layer by the weights array of size $[1*22]$. Figure (4-27) shows the final version of the output layer VI after many enhancements to reduce the consumed resources. Table (4-11) lists the consumed resources and the maximum operation frequency.

Device utilization	Before	After
Number of Slices(4656)	228%	81%
No. of slices FF(9312)	185%	63%
No. of 4 input LUTs(9312)	109%	53%
Minimum period	13.744ns	11.180ns
Maximum frequency	72.760MHz	89.443MHz

It is obvious from the last table (4-11), that the percentage of the total slices in the SPARTAN chip is to be consumed for the output layer.

Array functions were used to create and manipulate arrays. It is worth here to mention some of these array functions that have great role in the design [41]:-

1- Decimate 1D Array Function: it can partition the elements of an array into many arrays or output arrays by distributing its elements to the output arrays. Also, by resizing this function, additional output terminals can be added.

2- Building Array Function: It concatenates multiple arrays or appends elements to an n-dimensional array.

3- Replace Array Subset Function: It replaces an element or sub-array from an array at a point specified by the pointer in index.

4- Index Array Function: This function returns the element or sub-array of n-dimension array. The location of the element or the sub-array is specified by the pointer index.

Also, local variables can be assigned (for read and write) to one of the controls or the indicators on the front panel of a the VI. Accordingly, this let to use the same array as controller or indicator in different places.

After we built the NN in LabVIEW environments using the mathematics (linear Algebra) as shown in Fig (4-28), those functions couldn't be implemented on FPGA because they are not provided in LabVIEW FPGA environments.

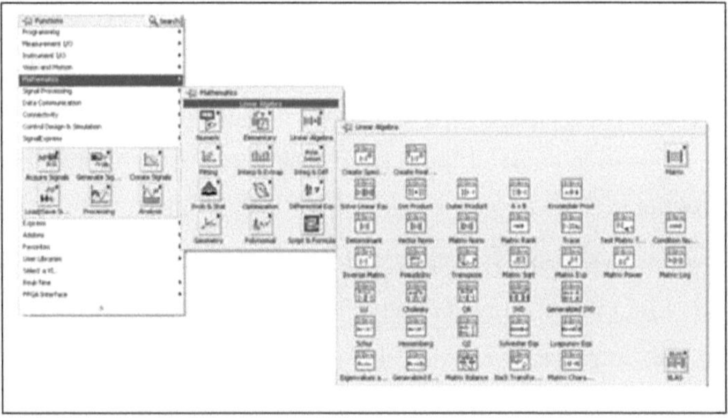

Figure (4-28):- The mathematics List in LabVIEW

4.6.5 Back-Propagation of Error:-

Here, the designer depends on back-propagation of error
algorithm steps from (step.6 to step.8) that listed in chapter two to
design a suitable one for our design as shown in Fig (4-29).

Firstly, the error information terms of each neural network output
units, that denoted by (Sk,Sj1,Sj2), were computed. These shown on
the first part of the figure (4-29). After that, as shown in the middle of
the same figure, the previous error information are used to compute
the bias correction terms (delw1,delv1,delv2,delvo1,delvo2) that can
be used to adjust the weights and biases of each output units. Finally,
each output units will update their bias and weights as shown in the
last part of the figure (4-29).

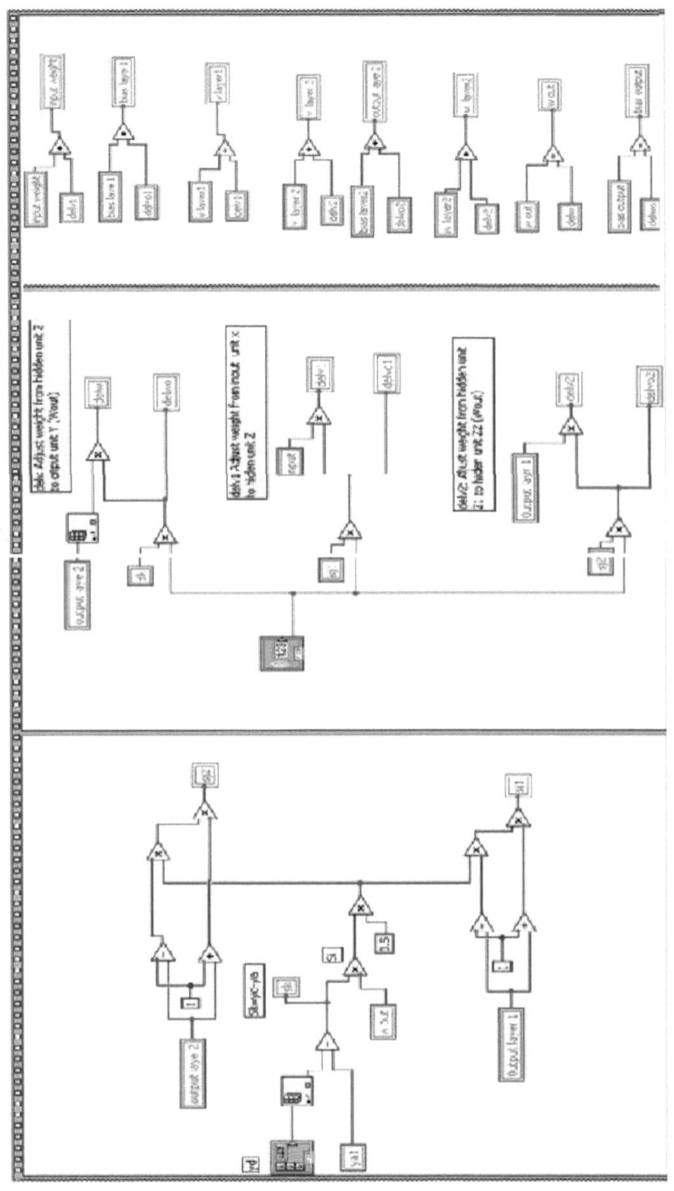

Figure (4-29): Back-propagation of error in LabVIEW

116

4.9.6 Neural Network Test

In order to test the designed and trained neural network on severe conditions the signal shown in figure (4-29a) was applied as a test input signal. The output of the network is shown in figure (4-29b).

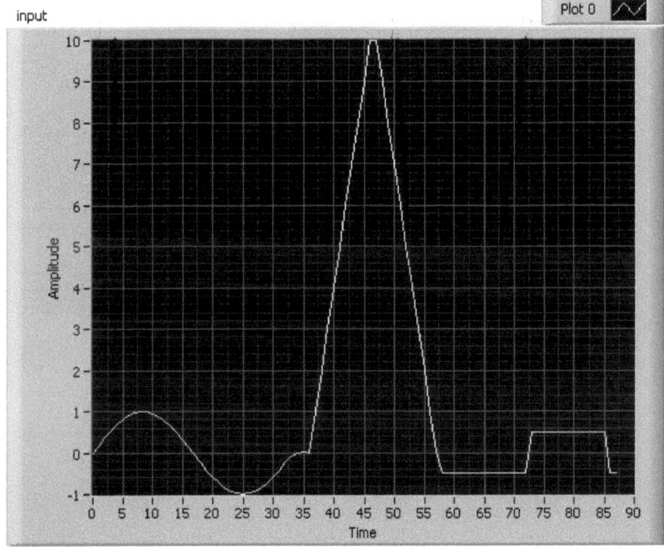

(a)

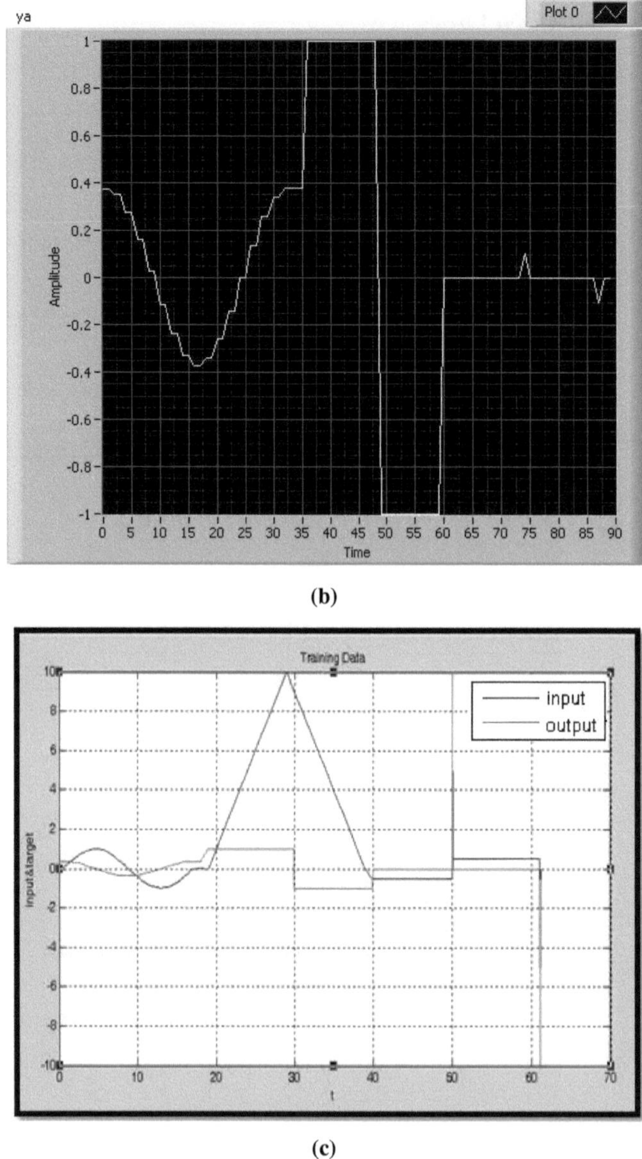

(b)

(c)

Figure (4-30): Many inputs waveforms in LabVIEW (a) training signals (patterns input to NN as well as RC circuit) (b) desired output after training the NN (c) Output signal of RC circuit.

4.11 <u>Methods Reducing FPGA Resources usage (hardware implementation optimization)</u>:-

After compiling, a report is generated which provides information about the speed and the size of the compiled VI. The Device Utilization Summary section provides information on the number of SLICEs used. This metric is the most important measure of the size of the program in hardware [6]. When the FPGA code needs to reduce the gate usage, National Instrument company recommends some of with the following steps which were followed throughout the work.

1- **Split up the FPGA code into parts:** To avoid large VIs that consume resources, the whole design was splited into part and for each part a separate FPGA program was created, mapped on this chip and tested. Each layer of the network was considered as individual part and the outputs of one layer are then fed as inputs to the next layer.

2- **Eliminate Arrays on the Front Panel:-**Arrays on the front panel take up extraordinary amounts of room on the chip as double buffering is required on the FPGA [40]. Therefore, lookup tables were used instead of array to reduce the FPGA resources usage.

3- **Minimize Front Panel Elements:** Every element on the Front Panel of an FPGA VI creates a register so that the host can use it to communicate. This process takes extra gates [40]. Here, local variables were used to avoid the increasing of the consumed elements on the front penal.

4- We used while loop which allowed using only one unit of certain mathematical function to process an array of data sequentially instead of using multiple parallel units to process the same data. Of course consumed resources are much reduced, but accordingly speed of processing decreases.

5- Data format and word length can affect reducing the consumed resources.

6- Sequential processing of Array elements. Figure (4-31) shows how to design processing Array of elements in LabVIEW FPGA, these elements processed by parallel units and therefore need more gates. If elements of the array are stored in shift registers, then processing them will be sequentially and therefore only one processing unit is required rather than (n-1) processing units that are used in parallel processing. By this method consumed resources are reduced to the half, but on the contrary, the processing speed decreases. Figure (4-32) shows the suggested design.

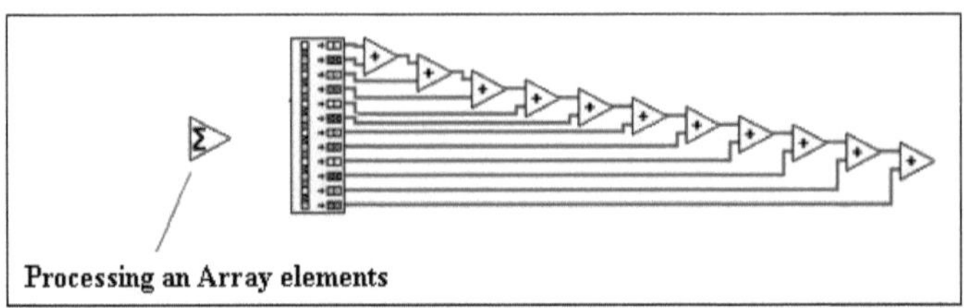

Figure (4-31): Parallel processing of Array elements in LabVIEW FPGA

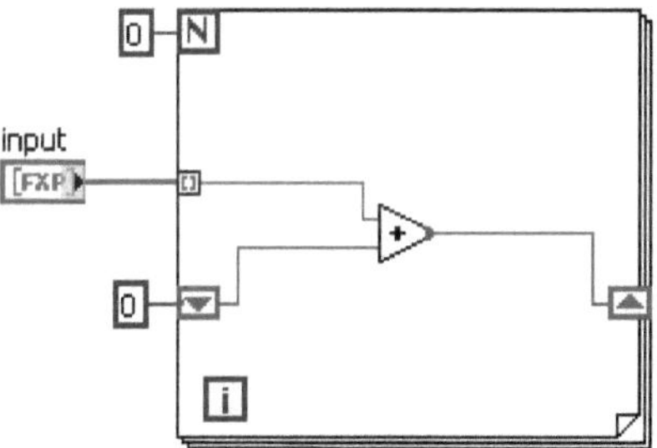

Figure (4-32): Sequential Processing of Array elements in LabVIEW FPGA

Following figures shows the hardware optimization of the RC circuit in LabVIEW FPGA.

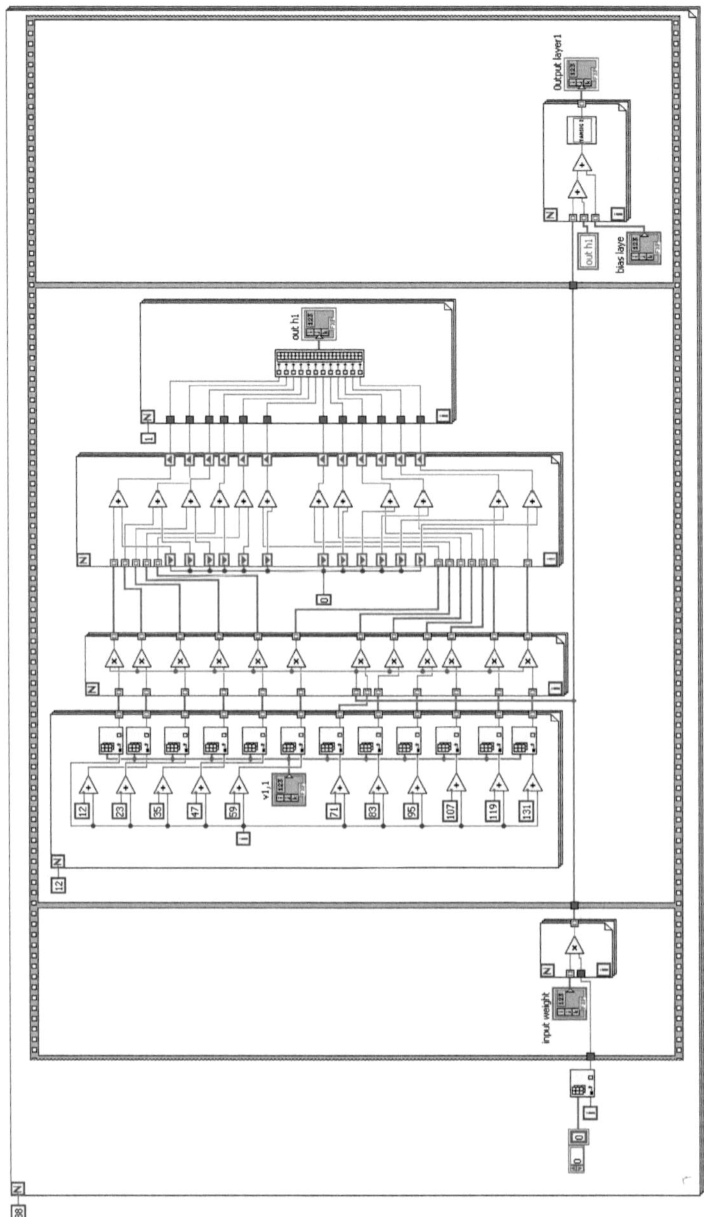

Figure(4.33): First recurrent layer

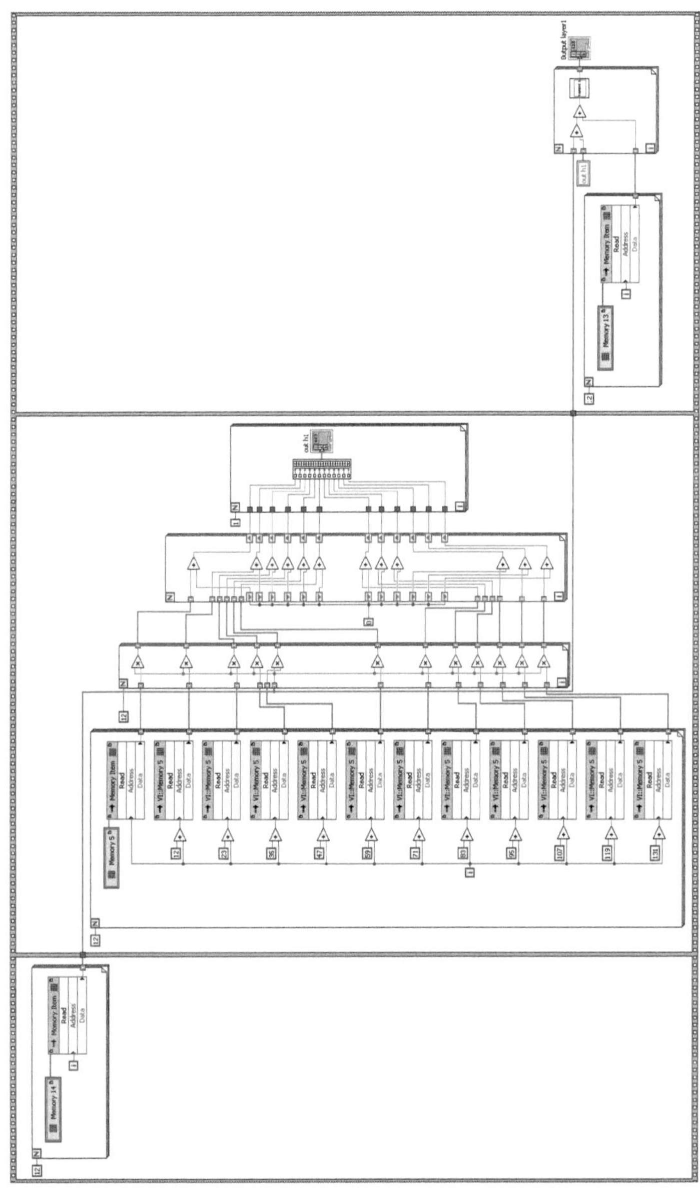

Figure(4.34): First layer with memory

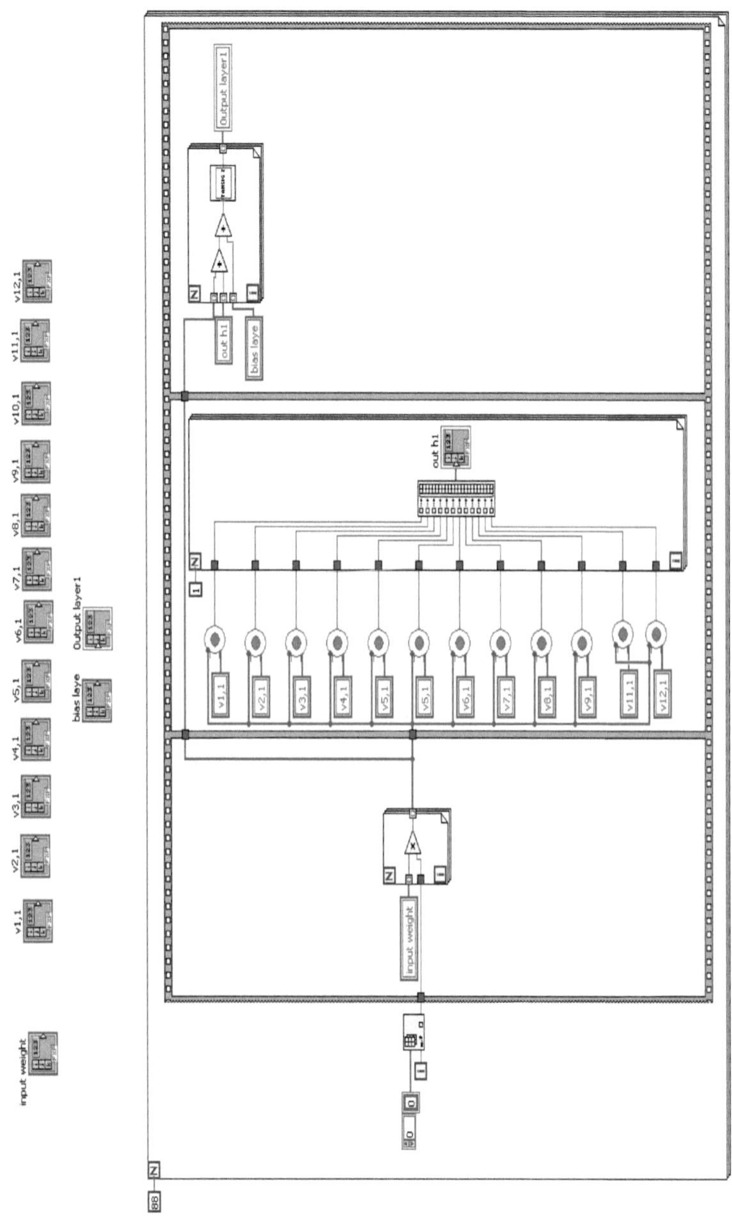

Figure(4.35): Input and first layer sub VI

Tables (4-12), (4-13), (4-14) shows the consumed resources of the first recurrent layer for the figures (4-33), (4-34), (4-35) respectively.

Table (4-12) : Consumed Resources of the Input and first layer

```
Device utilization summary:
---------------------------
Selected Device : 3s500eft256-5
 Number of Slices:              17187   out of  4656   369%  (*)
 Number of Slice Flip Flops:    22821   out of  9312   245%  (*)
 Number of 4 input LUTs:        22977   out of  9312   246%  (*)

Timing Summary:
---------------
Speed Grade: -5
   Minimum period: 11.680ns  (Maximum Frequency: 85.619MHz)
   Minimum input arrival time before clock: 3.316ns
```

Table (4-13) : Consumed Resources of the Input and first layer

```
Device utilization summary:
---------------------------
Selected Device : 3s500eft256-5
 Number of Slices:               9012   out of  4656   193%  (*)
 Number of Slice Flip Flops:    13869   out of  9312   148%  (*)
 Number of 4 input LUTs:        11165   out of  9312   119%  (*)

Timing Summary:
---------------
Speed Grade: -5
   Minimum period: 8.996ns  (Maximum Frequency: 111.157MHz)
   Minimum input arrival time before clock: 3.316ns
```

Table (4.14) : Consumed Resources of the Input and first layer

```
Device utilization summary:
---------------------------
Selected Device : 3s500eft256-5
 Number of Slices:              13775   out of  4656   295%
 Number of Slice Flip Flops:    18844   out of  9312   202%(*)
 Number of 4 input LUTs:        16307   out of  9312   175%(*)
Timing Summary:
---------------
Speed Grade: -5

   Minimum period: 11.520ns  (Maximum Frequency: 86.808MHz)
   Minimum input arrival time before clock: 3.316ns
```

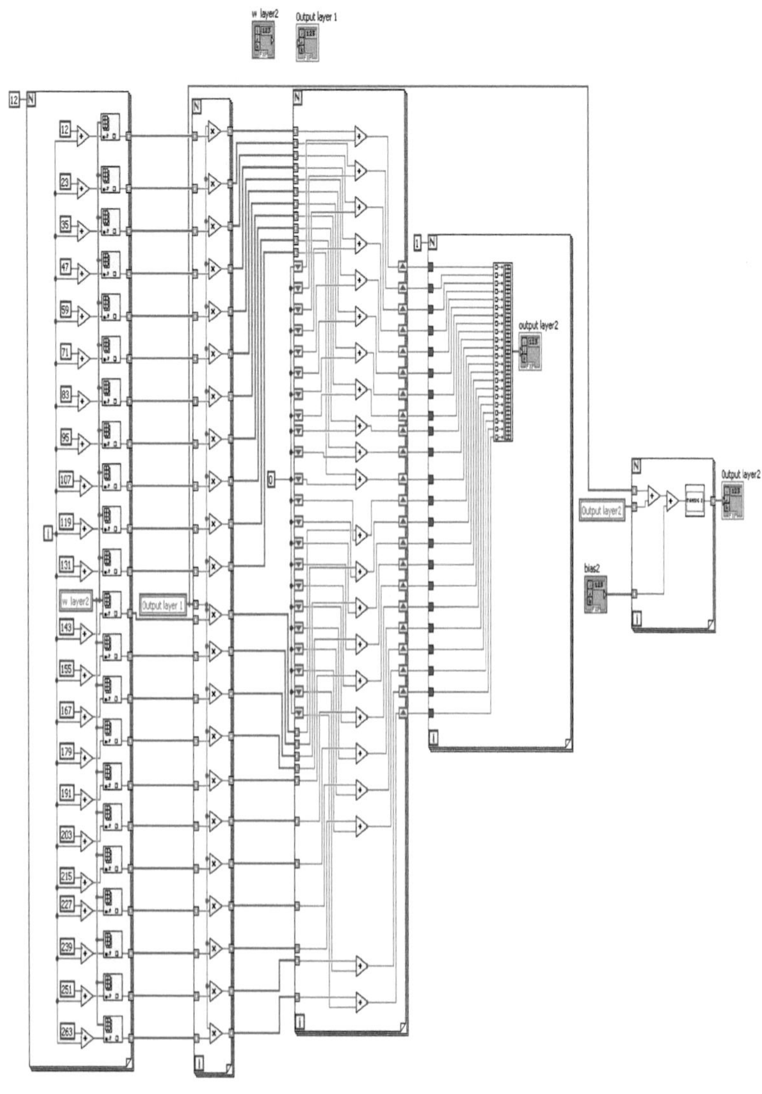

Figure (4.36) : Second layer

126

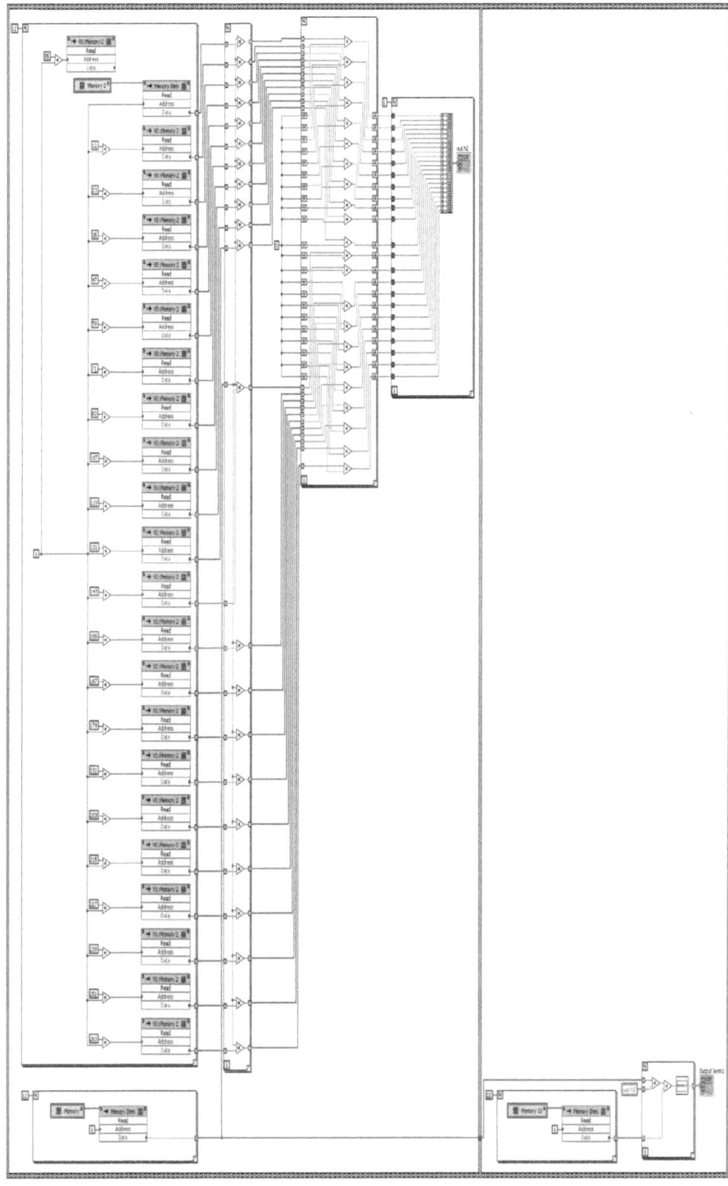

Figure (4.37): Second layer with memory

127

Consumed resources of second recurrent layer of the figures (4-36) and (4-37) can be seen in tables (4-15) and (4-16) respectively.

Table (4-15) : Consumed resources of second layer

```
Device utilization summary:
---------------------------

Selected Device : 3s500eft256-5

 Number of Slices:                  26624   out of 4656   571% (*)
 Number of Slice Flip Flops:        33698   out of 9312   361% (*)
 Number of 4 input LUTs:            40500   out of 9312 434% (*)
    Number used as logic:           40495

Timing Summary:
---------------

Speed Grade: -5
   Minimum period: 11.228ns  (Maximum Frequency: 89.063MHz)
   Minimum input arrival time before clock: 3.316ns
   Maximum output required time after clock: 0.871ns
```

Table (4-16) : Consumed Resources of the Second layer with memory

```
Device utilization summary:
---------------------------
Selected Device : 3s500eft256-5
 Number of Slices:                  15811   out of 4656 339% (*)
 Number of Slice Flip Flops:        24194   out of 9312 259% (*)
 Number of 4 input LUTs:            18155   out of 9312 194% (*)
    Number used as logic:           16542

Timing Summary:
---------------

Speed Grade: -5
   Minimum period: 11.228ns (Maximum Frequency: 89.063MHz)
   Minimum input arrival time before clock: 3.316ns
   Maximum output required time after clock: 0.871ns
```

CHAPTER FIVE

CONCLUSIONS AND FUTURE WORKS

5.1 <u>Conclusions</u>:-

In this thesis and from the results obtained, the following conclusions can be stated:

1- Firstly the designer use RNN (dynamic NN), which have dynamic memory, instead of feed forward or (static NN) because of these features:-

- Recurrent neural networks (RNNs) have dynamic memories, this feature make recurrent neural networks powerful tool for identification.
- Dynamic systems requires dynamic memories for dynamic mapping between the outputs and the inputs of that system.
- Recurrent neural networks RNNs requires less neurons and less computation time.
- Also, external noise has very little influence on recurrent neural networks (RNNs).

2- This work successfully presents a design of electronic circuit simulator which is a simple recurrent network (SRN).

3- The same network structure can be used to learn the input/output behaviour of any electronic circuit. The difference is only the weights of the neurons which are being fixed after training on the circuit under consideration.

4- By using NN to simulate electrical circuits, it can be used to detect faults and achieve compensation if there is any drift in the characteristics of the circuit.

5- The NN used in this work has two probes that can detect two points in any circuit. If it is required that the NN has to have more than two probes, another NN structure has to be designed. Mainly, other context layers are required.

6- The difference (i.e. error) between the output of the real electronic and the output of the neural network can be used in controlling the drift in the characteristic of the electronic circuits.

7- The neural network may fail to simulate the real electronic circuit, when the latter operates on frequency greater than the maximum frequency that the neural network can operate on.

5.2 <u>Future Works:-</u>

1- It is suggested for the future to train recurrent neural networks on more than two points of the electronic circuits.

2- In the future, we hope this work to be completed by adding A/D and D/A convertors respectively on the input and output of the network.

3- For the neural networks that simulated logical circuits, they are required to be tested on sequential circuits since they may need context layers because current states of sequential circuits depend on previous states as well current inputs.

References

[1] T. Callinan, **"Artificial Neural Network Identification and control of the inverted pendulum"**, technical report, August 2003.

[2] G. L. Dempsey, A. G. Pintoy and J. A. Wood. **"A New Design Strategy for the Tank and Hopfield Neural Analog-to-Digital Converter"**, Proceedings, 27th Asilomar Conference on Signals, Systems and Computers, Vol. No.1, pp.375-380, Pacific Grove, California, November 1993.

[3] S. WhanLee and H. HeonSong, **"A New Recurrent Neural-Network Architecture for Visual Pattern Recognition"**, IEEE Transactions On Neural Networks, Vol.8, NO.2, pp.331-340, March 1997.

[4] E.O.Dijk, **"Analysis of Recurrent Neural Networks with Application to Speaker Independent Phoneme Recognition"**, M.Sc. Thesis, University of Twente, department of Electrical Engineering, Enschede, The Netherlands, June 1999.

[5] P. Pogula, **"Developing Neural Network Applications Using LabVIEW"**, M.Sc. Thesis, University of Missouri, Columbia, JULY 2005.

[6] A. Kalinli and S. Sagiroglu, **"Elman Network with Embedded Memory for System Identification"**, Journal Of Information Science And Engineering 22, pp.1555-1568, 2006.

[7] Y.C. Cheng,W.Qi and J. Zhao, **"A New Elman Neural Network and Its Dynamic Properties"**, Cybernetics and Intelligent Systems IEEE Conference ,pp. 971 – 975, China ,2008.

[8] Z. Aimin, Z. Hang, L. Hong and C. Degui, **"A Recurrent Neural Networks Based Modeling Approach for Internal Circuits of Electronic Devices"**, in IEEE International

Symposium on   Electromagnetic Compatibility, Zurich, pp. 293-296,Zurich, 2009.

[9] J.A. Samath, P.S. Kumar and A. Begum, **"Solution of Linear Electrical Circuit Problem using Neural Networks"**, International Journal of Computer Applications (0975 – 8887), Vol. 2, No.1, pp.6-13, May 2010.

[10] Z.Zhang, **"Training Elman Neural Network For Dynamic System Identification Using An Adaptive Local Search Algorithm"**, International Journal of Innovative Computing, Information and Control (ICICInternational),Vol.6, No.5, pp.2233-2243, May 2010.

[11] T. Rashid, **"Shape Recognition through an Alternative Recurrent Network Architecture"**, Journal of Computer Science Informatics & electrical Engineering, Vol.2, Issue 1, 2010.

[12] G.Goos, J.Hartmanis and J.Leeuwen, **"Artificial Neural Networks"**, Springer-Verlag Berlin Heidelberg, New York , 1995.

[13] L. Fausett, **"Fundamentals Of Neural Networks"**, ISBN 10: 0133341860 / 0-13-334186-0,ISBN 13: 9780133341867.

[14] J.R.Rabuñal and J.Dorado, **"Artificial Neural Networks In Real Life Application"**, by Idea Group Inc., University of A Coruña, Spain, London, England, 2006.

[15] A. Saxena and A. Saad, **"Evolving an Artificial Neural Network Classifier for Condition Monitoring of Rotating Mechanical Systems"**, Journal of Applied Soft Computing, Elsevier Publishers,Vol.7, No.1 ,pp. 441–454, August 11 2005.

[16] U.C.Peñaloza, J.Esquer, B.Rios, **"A Methodology for Implementation of the Execution Phase of Artificial Neural**

Networks in Digital Hardware Devices", Electronics, Robotics and Automotive Mechanics Conference in IEEE, pp. 422-427 2008.

[17] H.BDemuth, M.T. Hagan and M.H. Beale, Users' Guide for the **"Neural Network Toolbox™ 7"**, Math Works, Inc., 1992–2010.

[18] H.KKwan, **"simple sigmoid_like activation function suitable for digital hardware implementation"**, Electronics letters, vol.28 No.15, pp. 1379-1380, 16th July 1992.

[19] R.A.Khalil, S. A. Al-Kazzaz, **"Digital Hardware Implementation of Artificial Neurons Models Using FPGA"**, Al-Rafidain Engineering, Department of Electrical Engineering, University of Mosul, Mosul, Iraq , Vol.17,No.2 ,April 2009.

[20] Y. H. Hu and J. Hwang, **"Handbook of NEURAL Network Signal Processing"**, by CRC Press LLC, London, England, 2002.

[21] C.T.Leonde, **"Neural Network Systems Techniques and Applications"**, Vol.5, **"Image Processing and Pattern Recognition"**, Academic Press ,San Diego, California, 1998.

[22] M.A.Arbib, **"The Handbook of Brain Theory And Neural Network"**, by Massachusetts Institute of Technology, London, England, 2003.

[23] C.T.Leonde, **"Neural Network Systems Techniques and Applications"**, Vol.6, **"Fuzzy Logic and Expert Systems Applications"**, Academic Press ,San Diego, California, 1998.
[24] C.T.Leonde, **"Neural Network Systems Techniques and Applications"**, Vol.7, **"Control and Dynamic Systems"**, **Academic Press"**, Academic Press , San Diego, California, 1998.

[25] H. Jaeger, **"Tutorial on training recurrent neural networks, covering BPPT,RTRL, EKF and the "echo state network"**

approach", German National Research Center for Information Technology, International University Bremen, GMD Report 159, 2002.

[26] C.T.Leonde, **"Neural Network Systems Techniques and Applications"**, Vol.3, Implementation Techniques, Academic Press, San Diego, California, 1998.

[27] Z. Zhang, Z. Tang, and C. Vairappan, **"A novel learning method for Elman neural network using local search"**, Neural Information Processing—Letters and Reviews, vol. 11, no. 8, pp. 181–188, 2007.

[28] A.Savran and S.Ünsal, **"Hardware Implementation Of a Feedforward Neural Network Using FPGAs"**, Ege University, Department of Electrical and Electronics Engineering, 2003.\

[29] Wikipedia–the free Encyclopedia, Wikipedia, May 2010, http://en.wikipedia.org/wiki., Wikimedia Foundation, Inc.

[30] Using the High Throughput Math Functions (FPGA Module), LabVIEW 2009 FPGA Module Help, National Instruments.mht Available at http://zone.ni.com/reference/en-XX/help/371599E-01/lvfpgaconcepts/using_ht_math/.

[31] NI Digital Electronics FPGA Board, National Instruments.mht Available at http://digital.ni.com/manuals.nsf/websearch/E9B5044C580D9D D2862575840068304A.

[32] Icon and Connector Pane, National Instruments.mht, Available at http://zone.ni.com/reference/enXX/help/371361F01/lvconcepts/intro_to_vis/

[33] Floating-Point Addition in LabVIEW FPGA, National Instruments.mht, Available at, http://Floating Point/Addition/in/LabVIEW/20FPGA//Developer/Zone/National /Instruments.mht

[34] Signal processing, Answers.mht, Available at http://www.answers.com/topic/signal-processi

[35] T. Koskela, M. Lehtokangas J. Saarinen and K. Kaski, " **Time Series Prediction With Multilayer Perceptron FIR and Elman Neural Network**", CiteSeerX, In Proceedings of the World Congress on Neural Networks, pp.491-496, 1996.

[36] J. M. Zurada," **Introduction to Artificial Neural Systems**", by West Publishing Company, New York, 1992.

[37] Y. Lee, "**A Neural Network Face Detector Design Using Bit-Width Reduced FPU In FPGA Reconfigurable Computing Platforms**", Saskatchewan, Canada, January 2007.

[38] A Non-Mathematical Introduction to Using Neural Networks, Heaton Research.mht, Available at http://www.heatonresearch.com/content/non-mathematical-introduction-using-neural-networks.

[39] National Instruments, LabVIEW User Manual, by National Instruments Corporation, May 2009, Available at http://digital.ni.com/manuals.nsf/websearch/109AC4061B4348D 68625749C00513D23.

[40] How Can I Optimize/Reduce FPGA Resource Usage and/or Increase Speed, National Instruments.mht, Available at http://digital.ni.com/public.nsf/allkb/311C18E2D635FA3386257 14700664816

[41] National Instruments, manual of LabVIEW 2009, LabVIEW 2009 Help, Available at http://digital.ni.com/manuals.nsf/websearch/47CB27B48CC2CB E786257603006D3491.

APPENDIX

Appendix A

Specifications of ELVIS

Specifications listed below are typical at 25°C unless otherwise noted.

FPGA	
FPGA	Xilinx XC3S500E-4FTG256C
System gates	500 k
Logic cells	10,476
Logic family	CMOS
Platform Flash configuration PROM	4 Mbit
Onboard USB-based FPGA/CPLD download/debug interface	
General-Purpose I/O	
GPIO lines	32 general-purpose digital I/O lines, 3.3 V, 8 mA maximum.
Analog Output	
Channels	4
Resolution	12 bits
DAC0, DAC1	0–3.3 V
DAC2, DAC3	0–2.5 V
Generation	Single point
Analog Input	
Channels	2
Resolution	12 bits, simultaneously sampled
Range	0–3.3V
Sample-and-hold acquisition time	39 ns
Acquisition	Singlepoint
General	
ON/OFF power switch	1

Reset Button	1
LEDs (discrete)	8
Slide switches	8
Push buttons	4
Seven-segment displays	2
Rotary encoder with push-button shaft	1
Clock oscillator	50MHz clock oscillator
12-pin expansion connectors (Pmod)	6, Digilent
Signal breadboard area	
For NI ELVIS	2
For FPGA	3
General-purpose breadboard area	1
NIELVIS connector interface	1, PCI type
Bus Interface	
USB	USB 2.0 Full-Speed
USB connector	Mini-USB Type B
Power	
DC power supply	15VDC, 650 mA

Power supplies
+15 V...1.5 A maximum1
–15 V...150 mA maximum1
+5 V..400 mA maximum1
Total combined power....................6 W maximum1

Physical

Dimensions	20.9 cm × 21.6 cm(8.25 in. × 8.5 in.)
Weight	284g (10 oz)

Maximum Working Voltage
Breadboard areas are only intended to be used for low voltage circuits (<42VAC, 60 VDC).

Appendix B

Hardware Components

Slide Switches:-The NI Digital Electronics FPGA Board has eight slide switches, SW0 through SW7, shown in Fig(B1) shows the circuitry of the slide switches. The switches typically exhibit about 2ms of mechanical bounce; there is no active debouncing circuitry. Switches have an output impedance of 2kΩ. When in the up, or ON, position, the switch connects the line to 3.3V, a logic High. When in the down, or OFF, position, the switch connects the line to ground, a logic Low.

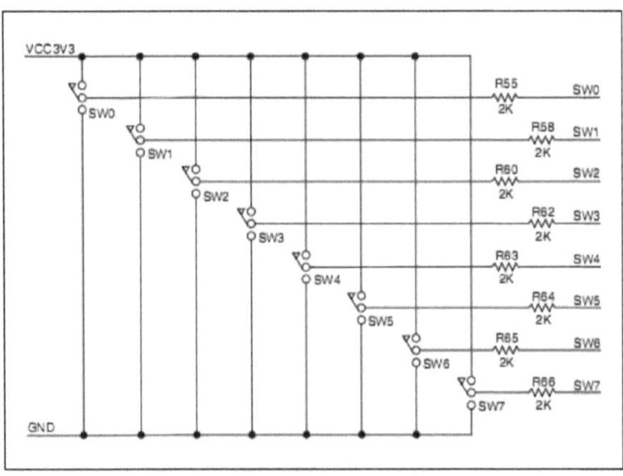

Figure (B1): Slide Switches Circuit Diagram

Push Buttons:-The NI Digital Electronics FPGA Board has four momentary-contact push-buttons, BTN0 through BTN3, shown in Fig (B3) shows the circuitry of the push buttons.

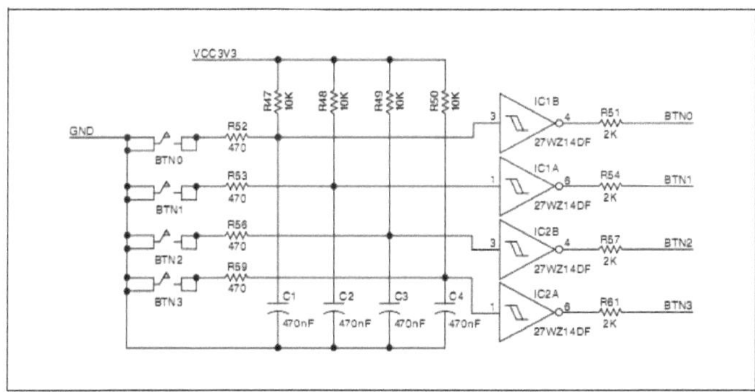

Figure (B3): Push Buttons Circuit Diagram

Pressing a push button connects a logic Low into an inverter, which outputs a logic High of 3.3 V into the associated line, as shown in Figure (B3). When the push button is not pressed, the power line goes into the inverter, which outputs a logic Low into the associated line. debouncing circuitry is implemented using a resistor and a capacitor on the push button signal line.

GPIO Lines:-The NI Digital Electronics FPGA Board has 32 general-purpose I/O lines, GPIO0 to GPIO31. Fig (B4).shows the circuitry of the GPIOlines.

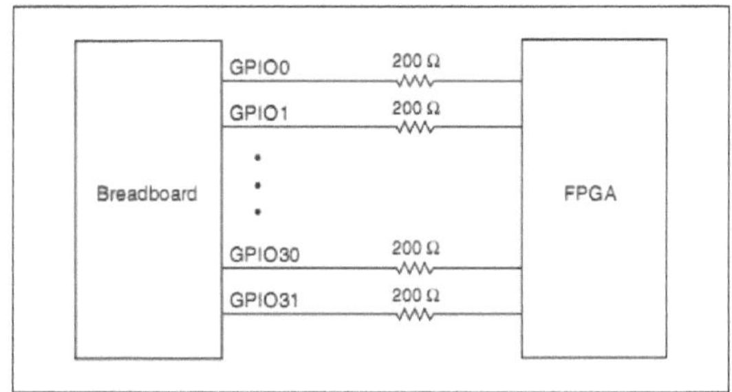

Figure (B4): GPIO Lines Circuit Diagram

Each GPIO line is connected to the FPGA through a 200 Ω current-limiting resistor [39].

Installation and Setup Instructions

To install and set up the NI Digital Electronics FPGA Board, complete the following steps.

1. Install the software you are going to use.
   a. Install NI LabVIEW as described in the LabVIEW Release Notes.
   b. Install the LabVIEW FPGA Module as described in the LabVIEW FPGA Module Release Notes.
   c. Install the NI Digital Electronics FPGA Board driver as described in NI Digital Electronics FPGA Board Driver Readme.
   d. (Optional) Install the NI ELVISmx software as described in the installation instructions on the software CD.

Or

   a. Install the Xilinx ISE software kit as described in the ISE Design Suite Release Notes and Installation Guide.

   b. Install the NI Digital Electronics FPGA Board driver as described in NI Digital Electronics FPGA Board Driver Readme.

2. Restart the PC if prompted.

3. Connect the NI Digital Electronics FPGA Board as described in one of the following sections:

- Stand-Alone Mode—To connect the NI Digital Electronics FPGA Board in stand-alone mode connected only to a PC, go to the Stand-Alone Mode section.
- NI ELVIS Mode—To connect the NI Digital Electronics FPGABoard in NI ELVIS mode, as an NI ELVIS prototyping board, goto the NIELVIS Mode section.

Stand-Alone Mode:-

The NI Digital Electronics FPGA Board can be used on a desktop as a stand-alone—or self-contained—device in stand-alone mode. The board requires a PC for new program download and optional application control/monitoring purposes. To install and set up the NI Digital Electronics FPGA Board in stand-alone mode, complete the following steps.

1. Connect the USB type A connector to the USB connector on the host PC.
2. Connect the USB type mini B connector to the NI Digital Electronics FPGA Board USB connector.
3. Connect the +15 VDC power adapter to the power connector on the NI Digital Electronics FPGA Board, then plug the power supply into
4. a wall outlet.

5. 4.Power on the NI Digital Electronics FPGA Board by moving the power

6. switch to the ON position.(Windows XP) Windows recognizes any newly installed device the first time the computer reboots after hardware is installed. On some Windows systems, the Found New Hardware Wizard opens with a dialog box for every NI device installed. Install the software automatically (Recommended) is selected by default. Click Next or Yes to install the software for the device and the USB cable ports. The green LD-G LED lights, indicating a good connection.

NIELVIS Mode

In NIELVIS mode, the NI Digital Electronics FPGA Board can be used as a prototyping board on an NIELVISII Series workstation. Install and set up the NI Digital Electronics FPGA Board and NI ELVIS II Series workstation as described in the Where to Start with NI ELVIS II Series document. Where the NI ELVIS II Series installation instructions refer to the prototyping board, complete the following instructions.

1. 1.Insert the NI Digital Electronics FPGA Board as a prototyping board, as described in the Where to Start with NI ELVIS II Series document.

2. Connect one end of the NI Digital Electronics FPGA Board USB cable to the NI Digital Electronics FPGA Board USB

connector, and the other end to the USB connector on the host PC.

3. Power on NIELVISII Series workstation.
4. Power on the NI Digital Electronics FPGA Board by moving the power switch to the ON position.

(Windows XP) Windows recognizes any newly installed device the first time the computer reboots after hardware is installed. On some

Windows systems, the Found New Hardware Wizard opens with

a dialog box for every NI device installed. Install the software

automatically (Recommended) is selected by default. Click Next or Yes to install the software for the device and the USB cable ports. The green LD-G LED lights, indicating a good connection.

Appendix C

Creating a Project and an FPGA Target VI

After making the important steps to work in Elvis mode, designer can make the following next steps that help him to start work in design his VI

1.LaunchLabVIEW as shown in Fig (C1).

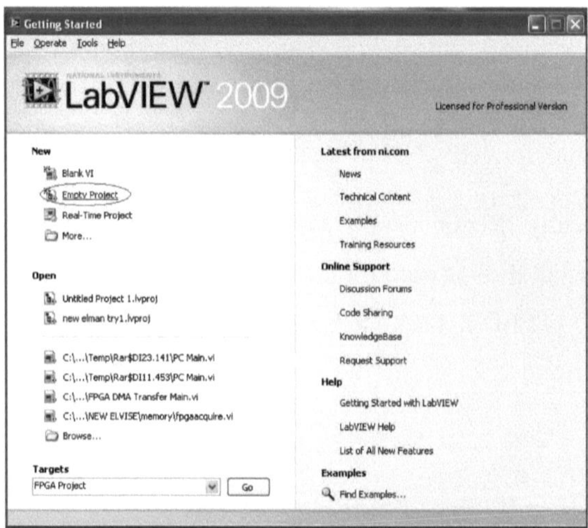

Figure (C1) Getting Stander window

2.In the Getting Started window, click Empty Project. The new project opens in the Project Explorer window as shown in Fig (C2).

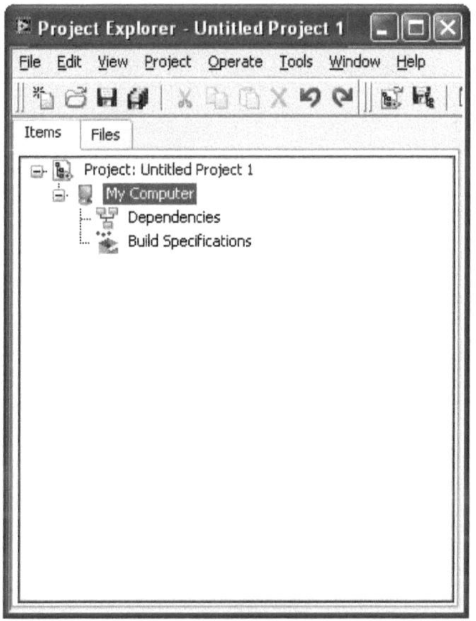

Figure (C2) Project Explorer window

3.Save the project as FPGA_Design .lvproj.

4.In the Project Explorer window, right-click My Computer and select New»Targets and Devices Fig (C3).

5.In the Add Targets and Devices on My Computer window, select New target or device, expand ELVIS, and highlight DE FPGA Board. Click OK. The target is discovered and the target and target properties are loaded into the project tree Fig (C4).

Figure (C3) Get new target

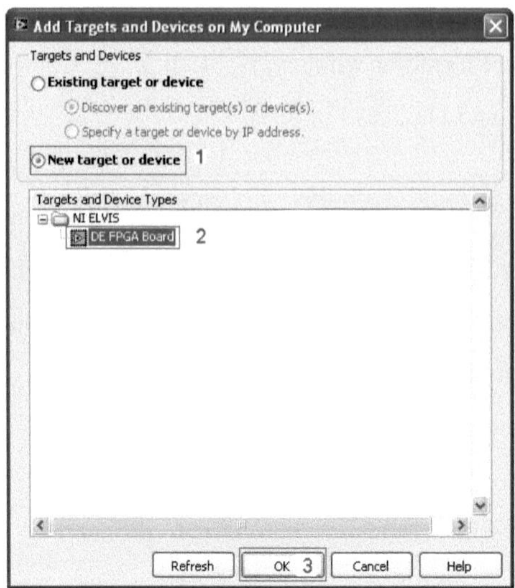

Figure (C4):Add target to project tree

6.In the Project Explorer window, right-click FPGA Target (Board1, DE FPGA Board) and select New» FPGAI/O. The New FPGA I/O window opens see Fig (C5).

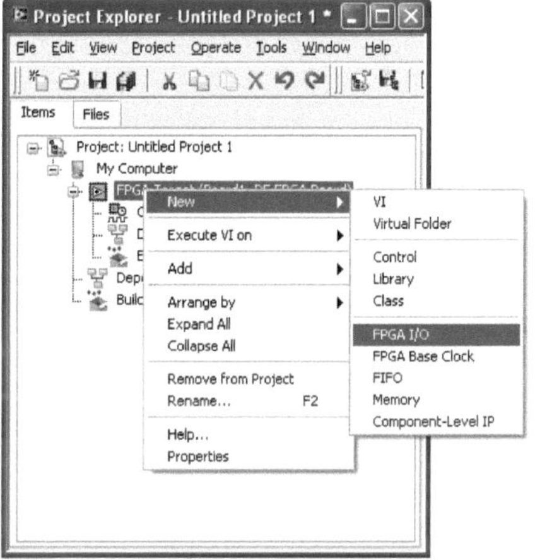

Figure (C5): FPGA (I/O)

7.In the Available Resources pane, expand Slide Switches and select SW0. Click Add to move SW0 to the New FPGA I/O pane. This adds this FPGA resource to the project.

8.Expand Push Buttons and select BTN0. Click Add to move BTN0 to the New FPGA I/O pane.

9.Expand LEDs and select LED0 and LED2. Click Add to move the LEDs to the New FPGA I/O pane as shown in Fig (C6).

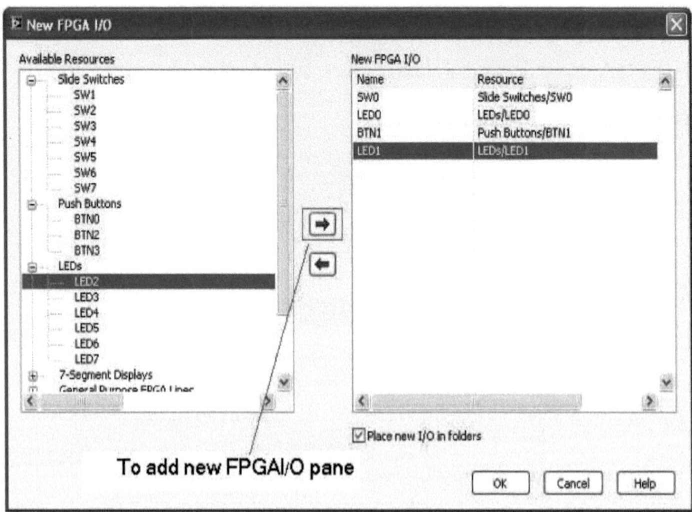

Figure (C6): Add New FPGAI/O

10.Click OK. Notice that the selected FPGA resources were added to the FPGA Target tree in the Project Explorer window, as shown in Fig (C7).

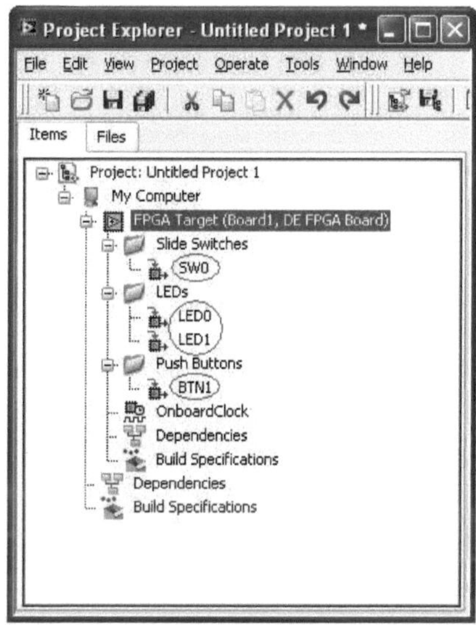

Figure (C7): FPGA Target Tree with New FPGA Resources

11.In the Project Explorer window, right-click FPGA Target (Board1,DE FPGA Board), and select New»VI. A blank VI opens. Select the block diagram window.

12.In the Project Explorer window FPGA Target (Board1, DE FPGA Board) tree view, select SW0 and LED0 and drag them onto the block Diagram, and also select BTN0 and LED2 and drag them onto the block Diagram. As shown in Figure Fig (C8)

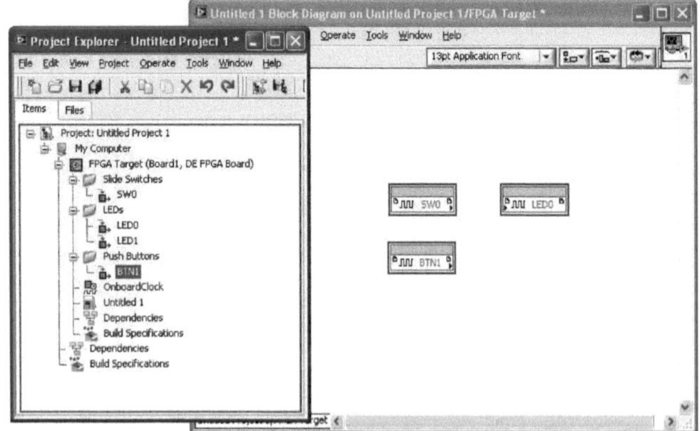

Figure(C8): Adding FPGA I/O to The Block Diagram

13.In the LabVIEW block diagram, wire SW0 output to the LED0 input. And wire BTN1 output to the LED1 input. And then add a While Loop around the resources. Show Fig (C9)

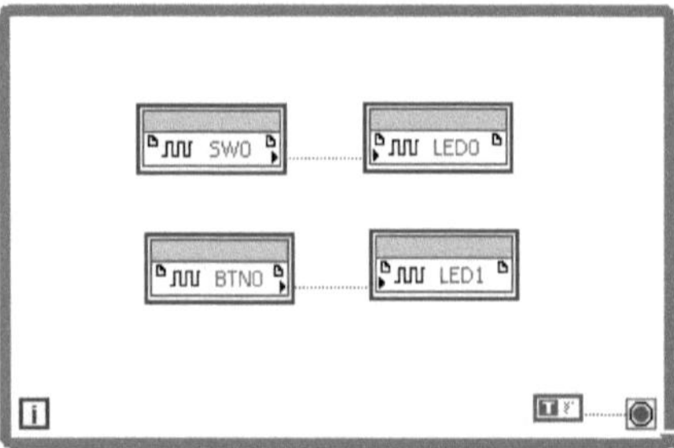

Figure (C9): Block Diagram

14. Save the VI as FPGA_Design.vi.

Appendix D

Programming with LabVIEW Software (Building a LabVIEW FGPA Design):

Programming the PROM:-

To change the default FPGA power-up application, the LabVIEW FPGA VImust be downloaded to the FPGA flash PROM by completing the following steps.

1. In the Project Explorer, right-click the target VI and select Download VI to Flash Memory.
2. LabVIEW displays the programming process. When the LabVIEW FPGA window displays *Download successful*, click OK.

Testing the Download

To test that the download was successful, complete the following steps.

1. Verify that switch SW9 is in the ROM position.

Fig (4-7). Switch SW9 in ROM Position.

Figure (4-7): Switch SW9 in ROM Position

2. Reboot the NI Digital Electronics FPGA Board by pressing the reset button.

3. Verify that the FPGA is running the (PROM) downloaded application.

Running the FPGA VI

After making the important steps to work in Elvis mode and creating project and an FPGA VI (Appendix B), designer can run his VI. The important steps of running the VIs that have been designed to work in FPGA. These steps

1. Verify that the USB cable is connected to the NI Digital Electronics FPGA Board and host PC, and the power switch is moved to the ON position.

2.Open the front panel of FPGA VI.

3.Click the Run button to run the VI. The application compiles VHDL code and generates a bit stream file that is downloaded into the FPGA configuration storage this shown in Fig (D1).

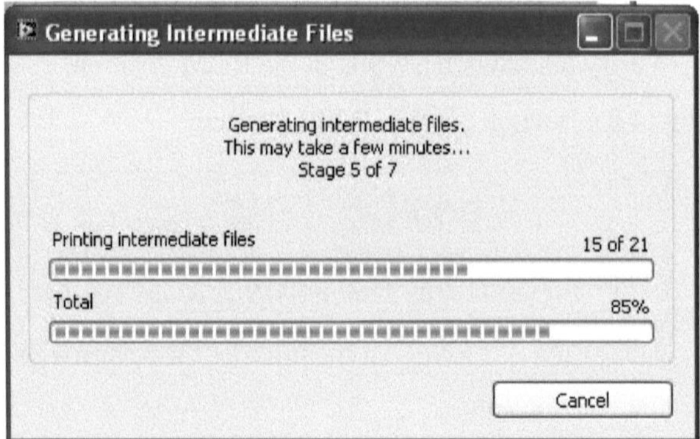

Figure(D1): Generate Intermediate Files

4- After that ,The Generating Intermediate Files window opens and displays the compilation progress. The LabVIEW FPGA Compile

Server window opens and runs as shown in Fig (D2). The compilation takes several minutes. The last one display the speed and the size for example

– IOBs : Input / Output Blocks

– MULT18X18SIOs : multipliers

– SLICEs : Combination of Look Up Tables (LUTs) and Flip Flops (FFs)

– BUFGMUXs :portal to the clock net, which is used to clock FFs.

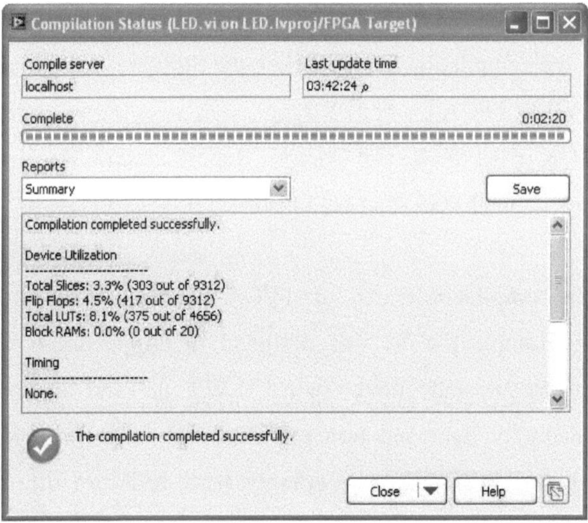

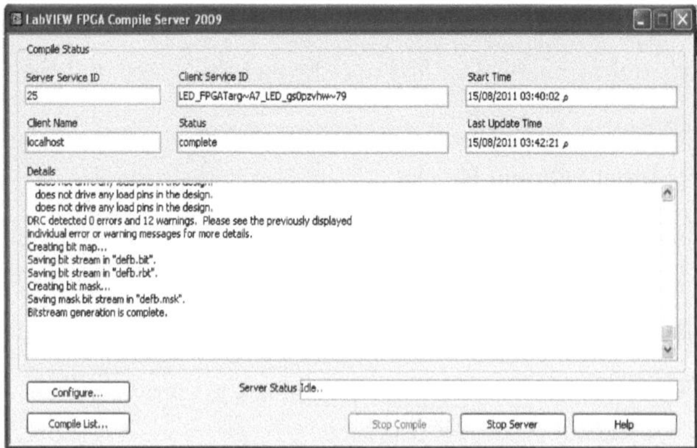

Figure(D2) Compile FPGA Code

5.When the compilation finishes, click the Stop Server button to close the LabVIEW FPGA Compile Server.

6.Click OK in the Successful Compile Report window.The application is running on the FPGA board at this time [39].

After the compilation is don, the FPGA work according to designed VI. For example the VI that designed in (Appendix B) and the designer try it and move switch SW0 up and down. LED0 correspondingly light and turn off. Also, when the designer Press button BTN0; LED2 correspondingly light and turn off. The last example shows that FPGA resources can be used and added directly in FPGA VI .

الملخّص

في هذه الرسالة طريقة جديدة لبناء معمارية الشبكات العصبية عن طريق اقتراح بناء هندسة الشبكة لبناء هندسة الشبكة العصبيةِ المعماريةِ.

إن شبكات الانتشار العكسي (Bp) لها محددات يمكن لها أن تتعلّم الإدخالات والإخراجات الثابتة. وعموماً فأنه من ناحية أخرى فان الشبكات العصبية المتكرّرة (RNNs) تَتطلّبُ أقل neuron في التركيب العصبي وأقل وقت حساب. فضلاً عن فإن احتمالية التعرض للضوضاء الخارجية تكون منخفضة. بسبب هذه الميزّات،وإن (RNNs) قد جذبت إنتباه الباحثين في مجال النظام الديناميكي. إذ أن شبكة Elman نوع من أنواع (RNNs) نماذج شبكة في الدوائر الكهربائيةِ أفضل من شبكات التوليد العصبية العكسية المتعددة الطبقة. شبكة Elman العصبية، التي هي نوع من أنواع الشبكة العصبية المتكرّرة وتسمى شبكات عصبية البسيطة أو المتكرّرة جزئياً. تركيب Elman بالإضافة إلى تخطيطه الساكن بين الادخال والاخراج ، وإن هنالك ربط داخلي للشبكة الذي يعمل كذاكرة مدى قريب وهي قادرة على تَمْثيل المعلوماتِ حول الادخالات السابقة. ومن هذه الفكرة فأن هذه الربط الداخلي يمكن ان يمثل الماضي.

تُبيّن النتائج بأن الخوارزمية المقترحة ENN تعطي النتائج المتوقعة في لتمييز الدوائر الالكترونية وبعد ذلك يمكن لهذه الاداة أن تكون أداة للسيطرة على الدائرة الالكترونية وعند استخدام شبكة أخرى كبديل عن تدريب شبكة (ENN) فانه الشبكة فشلت في إعطائنا النتائج المطلوبة.

LabVIEW برامجُ برمجة تخطيطية طوّرت National Instruments . تتركزت هذه الإطروحةِ على LabVIEW FPGA لبناءElman Neural Network . وبعدها هذا التصميم طُبّقَ على لوحةِ Elvis. Elvis لوحة LabVIEW FPGA أوصلتْ إلى محطة العمل الفرعيةِ بواسطة وصلات Elvis.

الإهداء

إلى معلمنا الأول إلى سيدي وسيد الخلق وشفيعي عند الله

سيدنا محمد صلى الله عليه وسلم

إلى من رباني على مكارم الأخلاق وزرع الثقة في نفسي وغرس في نفسي حب المعرفة

إلى من تمنيت أن يكون معي واسأل الله أن يتقبله شهيداً ويدخله فسيح جناته

أبي الحبيب . . . . جمعني الله به في عليين

إلى من جادت بكل ما تملك من عطف وحنان وسهر ورعاية ودعاء إلى جدول العطاء

الذي لا ينضب والنفس التي ما بخلت يوماً على أحد إلى النفس الكريمة

أمي الحبيبة . . . . رزقني الله برها

إلى من كان بعد الله عوني إلى من كان النبع الوفير في تزويدي بالعلم والمعرفة إلى الذي لم

يتوانى في توجيهي ونصحي

مشرفي . . . . بارك الله فيه

إليكم جميعاً اهدي بحثي هذا

مروة عز الدين

Printed by Books on Demand GmbH, Norderstedt / Germany